I0053522

Computational Simulation and Experimental Techniques for Nanofluid Flow

Edited by

Sabyasachi Mondal

Department of Mathematics
North- Eastern Hill University (NEHU)
Shillong, Meghalaya, India

Computational Simulation and Experimental Techniques for Nanofluid Flow

Editor: Sabyasachi Mondal

ISBN (Online): 978-981-5223-70-5

ISBN (Print): 978-981-5223-71-2

ISBN (Paperback): 978-981-5223-72-9

First published in 2024.

need for a court order if at any point you breach any terms of this License Agreement. In no event will any delay or failure by Bentham Science Publishers in enforcing your compliance with this License Agreement constitute a waiver of any of its rights.

3. You acknowledge that you have read this License Agreement, and agree to be bound by its terms and conditions. To the extent that any other terms and conditions presented on any website of Bentham Science Publishers conflict with, or are inconsistent with, the terms and conditions set out in this License Agreement, you acknowledge that the terms and conditions set out in this License Agreement shall prevail.

Bentham Science Publishers Pte. Ltd.
80 Robinson Road #02-00
Singapore 068898
Singapore
Email: subscriptions@benthamscience.net

BENTHAM
SCIENCE

CONTENTS

PREFACE

Advancements in the field of nanofluid flow have revolutionized numerous industries, ranging from energy and biomedical applications to electronics and manufacturing. The intricate behavior and unique properties of nanofluids have captured the imagination of researchers worldwide, sparking an intense exploration of their fundamental characteristics and practical applications. As a result, the need for comprehensive resources that bridge the gap between computational simulation and experimental techniques has become increasingly evident.

This edited volume, titled "Computational Simulation and Experimental Techniques for Nanofluid Flow," represents a collective effort of distinguished experts and researchers who have devoted themselves to unraveling the complexities of nanofluid dynamics. Drawing on their extensive experience and deep understanding of the subject, these contributors have compiled a rich compendium of knowledge that covers a wide spectrum of topics within this burgeoning field. The book is structured to provide a holistic overview of both computational and experimental methodologies, offering readers a balanced perspective on the subject matter. By exploring the synergies and complementarities between these two approaches, the editor and authors seek to foster a deeper appreciation for the challenges and possibilities that arise when investigating nanofluid flow phenomena.

The first section of the book delves into the theoretical foundations of nanofluid dynamics, providing a comprehensive framework to understand the principles underlying the behavior of nanoparticles in fluid environments. Various aspects of non-Newtonian behavior, hybrid nanofluids, and their daily life applications are discussed in the initial chapters, emphasizing the high heat transfer rates and enhanced thermal conductivity offered by hybrid nanomaterials. Furthermore, novel computational techniques such as the overlapping grid multi-domain bivariate spectral simple iteration method (OMD-BSSIM) are employed to solve hybrid nanofluid flow problems, utilizing the Chebyshev spectral collocation approach to handle the challenges of multi-domain problems.

Continuing with numerical investigations, the subsequent chapters explore the development of theoretical models and computational simulations for nanofluid flow phenomena. The third chapter focuses on the development of a theoretical model solved numerically using the Spectral Quasilinearization method. In the fourth chapter, the impinging oblique flow of nanofluids past a stretching sheet is analyzed, employing a cutting-edge Genetic Algorithm technique to understand the thermal characteristics and heat transport behavior over time. The fifth and sixth chapters concentrate on the behavior of Casson nanofluids and magneto-radiative nanofluids over different sheets, employing distinct numerical methods to solve the respective problems. The final two chapters investigate new nanofluid flow models and utilize the Spectral Quasi-Linearization Method (SQLM) to solve the governing equations, briefly discussing the influence of thermophoresis and Brownian motion.

Summarizing the discussions of all the chapters, the first section of the book establishes the theoretical foundations of nanofluid dynamics, providing a comprehensive framework for understanding the behavior of nanoparticles in fluid environments. Moving forward, the second section focuses on computational simulation techniques employed to study nanofluid flow, showcasing state-of-the-art methods such as MATLAB bvp4c, RK Method, cutting-edge Genetic Algorithm, Spectral Quasi-Linearization Method (SQLM), and the overlapping grid multi-domain bivariate spectral simple iteration method (OMD-BSSIM). These discussions shed light on the strengths and limitations of these methods in capturing the intricate behavior of nanofluids. The final section of the book addresses emerging trends and future directions in the field, exploring the potential impact of nanofluid flow on key industries and proposing avenues for further research and innovation.

As the editor, I aim to offer a comprehensive reference that stimulates the imagination, curiosity, and collaborative spirit of researchers, practitioners, and students embarking on their journey into the captivating world of nanofluid flow.

Sabyasachi Mondal
Department of Mathematics
North – Eastern Hill University (NEHU)
Shillong, Meghalaya, India

List of Contributors

Anindya Kundu	Department of Applied Mathematics, Maulana Abul Kalam Azad University of Technology, Haringhata, Nadia, West Bengal, India
Anwesha Dingal	Department of Applied Mathematics, Maulana Abul Kalam Azad University of Technology, Haringhata, Nadia, West Bengal, India
Debashis Mohanty	Department of Mathematics, C. V. Raman Global University, Bhubaneswar, India
Ganeswar Mahanta	Department of Mathematics, C. V. Raman Global University, Bhubaneswar, India
Gopinath Mandal	Siksha-Satra, Sriniketan, Visva-Bharati University, West Bengal, India
Hiranmoy Mondal	Department of Applied Mathematics, Maulana Abul Kalam Azad University of Technology, Haringhata, Nadia, West Bengal, India
Kamala Lochan Mahanta	Department of Mathematics, C. V. Raman Global University, Bhubaneswar, India
Nalini Kumar Sethy	Department of Mathematics, C. V. Raman Global University, Bhubaneswar, India
Nibedita Mandal	Department of Applied Mathematics, Maulana Abul Kalam Azad University of Technology, Haringhata, Nadia, West Bengal, India
Parveen Kumar	Department of Physics, Govt. P.G. Nehru College, Jhajjar, Haryana, India
Precious Sibanda	School of Mathematics, Statistics and Computer Science, University of KwaZulu-Natal, Pietermaritzburg, South Africa
Puspita Mondal	Department of Applied Mathematics, Maulana Abul Kalam Azad University of Technology, Haringhata, Nadia, West Bengal, India
Sachin Shaw	Department of Mathematics and Statistical Sciences, Botswana International University of Science and Technology, Private Bag 16, Palapye, Botswana
Salma Ahmedai	School of Mathematics, Statistics and Computer Science, University of KwaZulu-Natal, Pietermaritzburg, South Africa Faculty of Mathematical Sciences and Statistics, Al Neelain University, Khartoum, Sudan
Sabyasachi Mondal	Department of Mathematics, North- Eastern Hill University, (NEHU) Shillong, Meghalaya, India
Sewli Chatterjee	Department of Mathematics, Turku Hansda Lapsa Hemram Mahavidyalay, (Under Burdwan University), Mallarpur, Madian, West Bengal, India
Sharmistha Ghosh	Department of Applied Mathematics, Maulana Abul Kalam Azad University of Technology, Haringhata, Nadia, West Bengal, India
Sicelo Goqo	School of Mathematics, Statistics and Computer Science, University of KwaZulu-Natal, Pietermaritzburg, South Africa
Shilpa Taneja	Department of Physics, N.B.G.S.M. College, Sohna, Gurugram, Haryana, India
Uthman Rufai	School of Mathematics, Statistics and Computer Science, University of KwaZulu-Natal, Pietermaritzburg, South Africa
Vikas Poply	Department of Mathematics, KLP College Rewari, Haryana, India

CHAPTER 1

Cattaneo-Christov Heat Flux based Darcy-Forchheimer Hybrid Nanofluid Flow with Marangoni Convection above a Permeable Disk

Nalini Kumar Sethy[1], **Debashis Mohanty**[1], **Ganeswar Mahanta**[1], **Kamala Lochan Mahanta**[1] and **Sachin Shaw**[2,*]

[1]*Department of Mathematics, C. V. Raman Global University, Bhubaneswar, India*

[2]*Department of Mathematics and Statistical Sciences, Botswana International University of Science and Technology, Private Bag 16, Palapye, Botswana*

Abstract: The study highlights a rising number of fluids such as nanofluids and hybrid nanofluids encountered in daily life that exhibit non-Newtonian behavior and they are exploited in manufacturing due to their high heat transfer rate becoming more and more important as time goes on. The focus is on hybrid nanomaterials because they increase liquid alloys' and fluids' thermal conductivity. The various investigations on a thermal Marangoni convective flow of aluminium alloy and Boehmite alumina nanoparticles into gasoline oil in base fluid water on a steady Darcy-Forchheimer flow are covered. With the system's exponential heat generation and viscous dissipation, the thermal impact is more pronounced in the presence of Cattaneo-Christov heat flux. To simplify the highly coupled nonlinear governing equations (PDEs) and the boundary conditions (BCs), a suitable similarity conversion is being applied. The outcomes of the conversion equations and their BCs are evaluated by MATLAB bvp4c routine with the shooting techniques. Through the use of graphs and tables, it has been determined how different governing factors affect the velocity, temperature, skin friction, and Nusselt number. A quick comparison between a hybrid nanofluid and a nanofluid is displayed in each graph. We also discussed the system's Bejan numbers and entropy creation. It has been observed that the proportion of heat transmission increases with heat generation but decreases with the Marangoni number. A nonlinear increase in the permeability constant and Brinkman number results in a rise in entropy generation.

Keywords: Catteneo-Christov heat flux, Darcy-Forchheimer flow, Entropy generation, Hybrid nanofluid, Marangoni convection, Nanoflui.

[*] **Corresponding author Sachin Shaw:** Department of Mathematics and Statistical Sciences, Botswana International University of Science and Technology, Private Bag 16, Palapye, Botswana; E-mail: sachinshaw@gmail.com

Sabyasachi Mondal (Ed.)

INTRODUCTION

Nowadays, the study of nanofluid and hybrid nanofluid has become the hottest topic among researchers and scholars.

Several researchers conveyed their opinion regarding hybrid nanofluids with a high heat transmission rate. In order to speed up the rate of heat transmission, Choi *et al.* [1] first presented the idea of a nanofluid. The models of the hybrid nanofluid deal with new exciting possibilities to enhance the heat transfer of fluid and demonstrate the characteristics for future applications in the field of biochemical engineering, biomedical engineering, and in detergents other than non-Newtonian fluids. Recently Nayak *et al.* [2] examined entropy generation in hybrid fluid flow between two parallel plates. After that, many researchers deliberated on hybrid-fluidity experimentally and numerically [3–5]. The idea of a hybrid nanofluid is taken into consideration to aggregate two different nanoparticles to enhance thermal conductivity [6-9]. Suresh *et al.* [10, 11] introduced the concept of hybrid nanofluid and its concept provides a broad area for researchers to provide their worthy efforts. Baghbanzadeh *et al.* [12] also analyzed the accumulation of spherical silica nanotubes. Nine *et al.* [13] explored the Al_2O_3–MWCNTs and the recent study on hybrid nanofluid subjects to consider the thermal properties [14-21]. An inexhaustible and pollution-free solar radiation energy is utilized to fulfill the increasing demand [22]. The conjugate influence of entropy generation and thermal radiation over various geometries was studied by several researchers [23-28]. At present, concentrated solar collectors are used to provide energy for low- and medium-temperature processes, including improved absorption solar refrigeration systems [29], air cooling [30], and water heating [31]. In general, turbulent flows increase the fluid flow and heat transfer [32]. To propose better models for turbulent nanofluid flow, some researchers have compared them [33–35]. Nanofluids can use the current model for heat and mass transfer over different geometries [36–39].

Despite several works on Darcy-Forchheimer law in dissimilar non-Newtonian viscoelastic liquid inspected by countless investigators, for instance, Darcy-Forchheimer, viscous dissipation, gasoline oil, and biometer alumina were investigated by Mohanty *et al.* [40]. Physical characteristics of double chemically reacting hybrid nano-fluid flow over the stretched media dealing with MHD and Darcy Forchheimer law have been examined by Rashidi *et al.* [41]. Mahian *et al.* [42] and Goodarzi *et al.* [43] utilized Darcy Forchheimer's law in a variable tricked surface of the disk and flow across a curved surface. Siddiqui *et al.* [44] probed a rotating and stretched flow of bi-convective nanofluids subject to Darcy

Forchheimer law. The significance of Darcy Forchheimer's law on the dynamics of magnetized nanofluid in a circular form has been examined by Ali *et al.* [45].

In the current investigations, we explored the effect of Darcy-Forchheimer's theory's influence on the porous rotating disk with magnetic field, thermal radiations, and accumulations of Boehmite alumina into gasoline oil. The primary objectives of this work are (a) To analyze the influence of the Marangoni convection for a Darcy-Forchheimer hybrid nanofluid flow across a stretching disk, (b) Heat generation of the system with Catteneo-Christov heat flux and non-linear thermal radiation, (c) The entropy of the system to understand heal losses, and (d) To carry out an assessment of general and hybrid nanofluid on the system and entropy. The effort presented is new as per the literature assessment. Developed partial differential equations (PDEs) demonstrate that the fluid flow is changed into a set of ODEs by applying a resemblance conversion. A numerical procedure namely BVP4C is executed to find results. Attained solutions are evaluated and concluded in tables and graphs.

Terminology

(u, w)	Velocity components
ρ_{nf}	Density of nanofluid
$(\rho C_p)_{nf}$	Heat capacitance of nanofluid
$(\rho C_p)_f$	Thermal capacity of the underlying fluid
$(\rho C_p)_s$	Heat capacity of solid nanoparticles
ρ_s	Density of solid nanoparticles
ρ_f	Fluid's base density
μ_{nf}	Dynamic viscosity of nanofluids

(Box) cont.....

μ_f	Dynamic viscosity of basefluid		
k_{nf}	Thermal conductivity of nanofluid		
k_f	Thermal conductivity of base fluid		
k_s	Thermal conductivityof nanoparticle		
σ_{nf}	Electrical conductivity of nanofluid		
σ_s	Electrical conductivity of nanoparticles		
σ_f	Electrical conductivity of base fluid		
T_w	Surface temperature		
T	Fluid temperature	T_∞	Ambient fluid's temperature
k^*	Absorption coefficient	σ^*	Stefan Boltzmann constant
λ	Porosity variable		
K_p	Permeability constant	Ω	Dimensional constant
α_f	Thermal diffusivity of base fluid	Br	Brinkman number
Rd	Radiation parameter	β	Dimensional constant
K^*	Permeability of porous medium	ϕ	Solid volume fraction
P_r	Prandtl number	Rd	Radiation parameter
R_T	Thermal relaxation parameter		
Re_r	Reynolds number	Ma	Marangoni number
E_c	Eckert number	Q	Heat generation parameter

(Box) cont.....

C_b	Drag coefficient	θ_w	Temperature ratio parameter
F_r	Inertia coefficient	$F\left(=\dfrac{rC_b}{\sqrt{K_p}}\right)$	Non-uniform inertia coefficient

PROBLEM IDENTIFICATION

Let us consider a thermally convective Marangoni hybrid nanofluid flow of under the effect of an infinite porous disc, resulting in a continuous Darcy-Forchheimer flow (Fig. **1**). The nanoparticles are taken by the accumulation of Boehmite alumina $(\gamma - ALOOH)$ and aluminum alloy $(AA7072)$ into water. The Darcy-Forchheimer model's expression is elaborated by a viscous in-compressible fluid that neutralises the porous region. The system includes entropy formation in relation to viscous dissipation and non-linear thermal radiation. At z = 0, the disk's constant spin angular velocity Ω is taken into account. (u, v) are velocity components aligned in the corresponding directions (r, z).

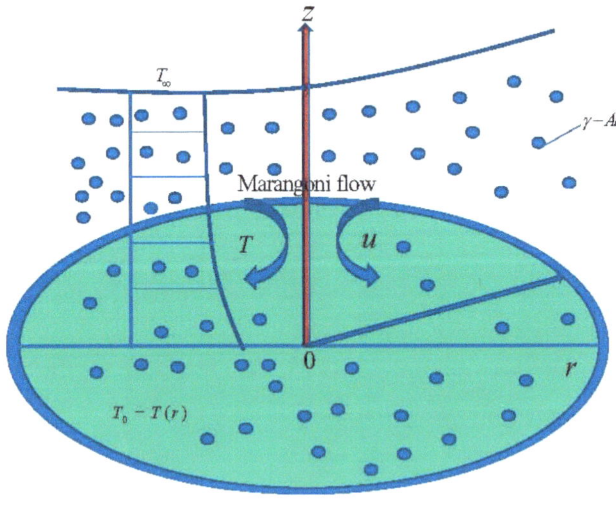

Fig. (1). Schematic flow diagram.

The flow equations are stated as:

$$\frac{\partial u}{\partial r} + \frac{u}{r} + \frac{\partial w}{\partial z} = 0, \tag{1}$$

$$\left(u\frac{\partial u}{\partial r}+w\frac{\partial u}{\partial z}\right)=\frac{\mu_{hnf}}{\rho_{hnf}}\frac{\partial^2 u}{\partial z^2}-\frac{\mu_{hnf}}{\rho_{hnf}K_p}u-\frac{C_b}{\sqrt{K_p}}u^2-\frac{\sigma_{hnf}}{\rho_{hnf}}B_0^2u, \quad (2)$$

With the decrease in heat, thermal diffusion is explained by the Cattaneo-Christov heat flux hypothesis as,

$$q+\lambda_T\left[\frac{\partial q}{\partial t}+V.\nabla q-q.\nabla V+(\nabla.V)q\right]=-k_{hnf}\nabla T, \quad (3)$$

where q is denoted as heat flux. Velocity $V=(u,w)$ acts in the direction r and z. k_{hnf} is characterized as the hybrid nanofluid based on thermal conductivity, and λ_T signifies thermal relaxation time. Eq. (3) reduces to Fourier law and given,

$$q+\lambda_T[V.\nabla q-q.\nabla V]=-k_{hnf}\nabla T. \quad (4)$$

By utilising the aforementioned ideas, the problem's temperature equation is

$$\left(u\frac{\partial T}{\partial r}+w\frac{\partial T}{\partial z}\right)+\lambda_T\left[\begin{array}{l}u^2\frac{\partial^2 T}{\partial r^2}+w^2\frac{\partial^2 T}{\partial z^2}+2uw\frac{\partial^2 T}{\partial r\partial z}\\+\left(u\frac{\partial u}{\partial r}+w\frac{\partial u}{\partial z}\right)\frac{\partial T}{\partial r}+\left(u\frac{\partial w}{\partial r}+w\frac{\partial w}{\partial z}\right)\frac{\partial T}{\partial z}\end{array}\right]$$
$$=\left(\frac{k_{hnf}}{(\rho C_p)_{hnf}}+\frac{16\sigma^* T^3}{3k^*(\rho c_p)_{hnf}}\right)\left(\frac{\partial^2 T}{\partial r^2}+\frac{1}{r}\frac{\partial T}{\partial r}+\frac{\partial^2 T}{\partial z^2}\right)$$
$$+Q_0(T-T_\infty)\exp\left(-n\sqrt{\frac{\Omega}{v_f}}z\right)+$$
$$+\frac{\mu_{hnf}}{(\rho C_p)_{hnf}}\left[\begin{array}{c}2\left\{\left(\frac{\partial u}{\partial r}\right)^2+\left(\frac{u}{r}\right)^2+\left(\frac{\partial w}{\partial z}\right)^2\right\}\\+\left(\frac{\partial u}{\partial z}+\frac{\partial w}{\partial r}\right)^2\end{array}\right], \quad (5)$$

with suitable boundary conditions:

$$\mu_{hnf}\left.\frac{\partial u}{\partial z}\right|_{Z=0}=\left.\frac{\partial\sigma}{\partial z}\right|_{Z=0}=\left.\frac{\partial\sigma}{\partial T}\frac{\partial T}{\partial r}\right|_{Z=0}$$

$$T\big|_{Z-0} = T_0 = T_\infty + \beta r^2, T\big|_{Z\to\infty} = T_{\infty}, \tag{6}$$

where

$$\sigma = \sigma_0 \left\lfloor (1 - \gamma_T (T - T_\infty)) \right\rfloor, \text{ and } \gamma_T = \frac{-1}{\sigma_0} \frac{\partial \sigma}{\partial T}\bigg|_{T=T_\infty}.$$

The physical features of the hybrid nanofluid and the applied models are defined as:

Density $$\rho_{hnf} = (1-\phi)\rho_{bf} + \phi_1\rho_{s1} + \phi_2\rho_{s2} \tag{7}$$

Heat capacitance $$(\rho C_p)_{hnf} = (1-\phi_1)(1-\phi_2)(\rho C_p)_{bf} + \phi_1(\rho C_p)_{s1} + \phi_2(\rho C_p)_{s2} \tag{8}$$

Thermal diffusivity $$\alpha_{hnf} = \frac{k_{hnf}}{(\rho C_p)_{hnf}}, \tag{9}$$

Dynamic viscosity $$\frac{\mu_{hnf}}{\mu_{bf}} = \frac{1}{(1-(\phi_1+\phi_2))^{2.5}}, \tag{10}$$

and the hybrid nanofluid and thermal conductivity is considered as:

$$\begin{gathered}\frac{k_{hnf}}{k_{bf}} = \frac{k_{s1} + 2k_{bf}(s-1) - \phi_1(s-1)(k_{bf} - k_{s1})}{\phi_2(k_{bf} - k_{s2}) + k_{bf}(s-1) + k_{s2}}, \\ \frac{k_{bf}}{k_f} = \frac{k_{s1} + 2k_f(s-1) - \phi_1(s-1)(k_{bf} - k_{s1})}{\phi_2(k_{bf} - k_{s2}) + k_{bf}(s-1) + k_{s2}},\end{gathered} \tag{11}$$

$$\begin{gathered}\frac{\sigma_{hnf}}{\sigma_{bf}} = \frac{\sigma_{s1} + 2\sigma_{bf}(s-1) - \phi_1(s-1)(\sigma_{bf} - \sigma_{s1})}{\phi_2(\sigma_{bf} - \sigma_{s2}) + \sigma_{bf}(s-1) + \sigma_{s2}}, \\ \frac{\sigma_{bf}}{\sigma_f} = \frac{\sigma_{s1} + 2\sigma_f(s-1) - \phi_1(s-1)(\sigma_{bf} - \sigma_{s1})}{\phi_2(\sigma_{bf} - \sigma_{s2}) + \sigma_{bf}(s-1) + \sigma_{s2}}.\end{gathered} \tag{12}$$

The conversions are made for the current model as follows:

$$\left.\begin{aligned}(u,w)&=\left(r\Omega F(\eta),-\sqrt{\Omega\upsilon_f}\,H(\eta)\right),\\(T)&=\left(T_\infty+\beta r^2\theta(\eta)\right),\\\eta&=\sqrt{\frac{\Omega}{\upsilon_f}}\,z,\end{aligned}\right\}, \tag{13}$$

where η is the non-dimensional space along the rotatiting axis. Further, F, H, θ and ϕ are mapping of η. With ns (7-13), Eqs. (1), (2), (5) and (6) have the following format.

$$2F+H'=0, \tag{14}$$

$$\frac{1}{\varepsilon_1\varepsilon_2}\left(2F''-\lambda F\right)-F^2+HF'+(1+F_r)F'^2+\varepsilon_4 MF=0, \tag{15}$$

$$\begin{aligned}&\left\{\frac{k_{hnf}}{k_f}+\frac{4}{3}Rd\left[1+(\theta_w-1)\theta\right]^3\right\}\theta''+4\frac{Rd}{\Pr}\left[1+(\theta_w-1)\theta\right]^2(\theta_w-1)(\theta')^2\\&+\Pr\varepsilon_3F\theta'-2\Pr R_T\left(F^2\theta+F'H\theta'\right)+Q_T\,\theta\exp(-n\eta)+\frac{Br}{\varepsilon_1\varepsilon_2}\left(F'^2+H'^2\right)=0,\end{aligned} \tag{16}$$

$$\left.\begin{aligned}&F'(\eta)\big|_{\eta-0}=-2Ma\varepsilon_1, H(\eta)\big|_{\eta-0}=0, \theta(\eta)\big|_{\eta-0}=1,\\&F'(\eta)\big|_{\eta-\infty}=0, \theta(\eta)\big|_{\eta-\infty}=0,\end{aligned}\right\} \tag{17}$$

Where F_r represents the non-uniform inertia coefficient, Ma is the Marangoni number, λ is the porosity variable, K_p is the permeability parameter, Ω is the dimensional constant, Ec is the Eckert number, Rd represents radiation parameter, R_T is stated as thermal relaxation parameter, Re_r is the local Reynolds number, θ_w is the temperature ratio variable, $\Pr$ is the Prandtl number, Q_T is the heat generation parameter, Br is the Brinkman number, and written as:

$$\left.\begin{aligned}
&Ma=\frac{\beta\gamma_T}{\mu_f\Omega}\sqrt{\frac{\Omega}{\upsilon_f}},\mathrm{Re}_r=\left(\frac{r^2\Omega}{\upsilon_f}\right),\ F_r=\frac{rC_b}{\sqrt{K_p}},\mathrm{M}=\frac{\sigma_f B_0^2}{\rho_f\Omega},\ Ec=\frac{r^2\Omega}{\beta\left(C_p\right)_f},\\
&Q_T=\frac{Q_0}{\Omega\left(\rho c_p\right)_f},\theta_w=\frac{T_0}{T_\infty},\lambda=\frac{\upsilon_f}{\Omega K_p},\mathrm{Pr}=\frac{(\mu C_P)_f}{K_f},Br=\mathrm{Pr}\ Ec,\\
&Rd=\frac{4\sigma^* T_\infty^3}{k^* k_f},\ R_T=\Omega\lambda_T.
\end{aligned}\right\} \tag{18}$$

$$\varepsilon_1=\frac{\mu_f}{\mu_{hnf}},\ \varepsilon_2=\frac{\rho_{hnf}}{\rho_f},\ \varepsilon_3=\frac{\left(\rho c_p\right)_{hnf}}{\left(\rho c_p\right)_f},\ \varepsilon_4=\frac{\sigma_{hnf}}{\sigma_f}, \tag{19}$$

The dimensionless local skin friction coefficient is stated as:

$$\left(\mathrm{Re}_r\right)^{1/2}C_{fr}=\frac{1}{\varepsilon_1}\left\{\sqrt{\left[F'(0)\right]^2+\left[H'(0)\right]^2}\right\}. \tag{20}$$

The non-dimensional local Nusselt number $\left(\mathrm{Re}_r\right)^{-1/2}Nu_r$ reads

$$\left(\mathrm{Re}_r\right)^{-1/2}Nu_r=-\left(\frac{k_{hnf}}{k_f}+\frac{4}{3}Rd\left\{1+(\theta_w-1)\theta(0)\right\}^3\right)\theta'(0). \tag{21}$$

ANALYSIS OF ENTROPY

The local entropy generation of the nanofluids is specified by:

$$\begin{aligned}
\dot{S}_G=&\underbrace{\frac{k_{nf}}{T_\infty^2}\left[\frac{k_{hnf}}{k_f}+\frac{4}{3}Rd\right]\left(\frac{\partial T}{\partial z}\right)^2}_{\text{Thermal irreversibility}}\\
&+\underbrace{\frac{\mu_{nf}}{T_\infty}\left\{2\left[\left(\frac{\partial u}{\partial r}\right)^2+\frac{1}{r^2}u^2+\left(\frac{\partial w}{\partial z}\right)^2\right]+\left(\frac{\partial u}{\partial z}\right)^2+\left(\frac{\partial w}{\partial r}\right)^2\right\}}_{\text{Fluid friction irreversibility}}\\
&+\underbrace{\frac{\mu_{hnf}}{T_\infty K}\left(u^2+w^2\right)}_{\text{Fluid friction irreversibility}}.
\end{aligned} \tag{22}$$

Next, the right-hand side of Eq. (22) demonstrates that irreversible heat transfer is responsible for the first term's entropy creation, whereas irreversible fluid friction is responsible for the second term. The ratio involving the actual $\dot{S}_G$ and characteristic entropy rates $\dot{S}_0$ is represented by the entropy number, which is the entropy generation rate's dimensionless version.

Utilizing Eq. (13) to dimensionless outcomes of the local entropy generation provided in Eq. (22) and then with simplification, the entropy generation number N_G becomes:

$$N_G = \left(\frac{k_{hnf}}{k_f} + \frac{4}{3}Rd\right)\{\theta'(\eta)\} + B_r\left[\frac{4}{\mathrm{Re}_r}\{F(\eta)\}^2 + \left(\{F'(\eta)\}^2 + \{H'(\eta)\}^2\right)\right] + B_r K\{F(\eta)\}^2 + \{H(\eta)\}^2, \quad \textbf{(23)}$$

with $B_r = \dfrac{\mu_{nf}\Omega v_f}{k_{nf}\beta^2 r^4}$ is the Brinkman number, $\mathrm{Re}_r = \dfrac{\Omega r^2}{\upsilon_f}$ is the Reynolds rotational number, and $N_G = \dfrac{\dot{S}_G}{\left(\dfrac{k_{nf}}{T_\infty^{\,2}}\dfrac{\Omega}{v_f}\beta^2 r^4\right)}$ is the dimensionless entropy generation rate.

The Bejan number is expressed as follows:

$$Be = \frac{\left(\dfrac{k_{hnf}}{k_f} + \dfrac{4}{3}Rd\right)\{\theta'(\eta)\}^2}{\left(\dfrac{k_{hnf}}{k_f} + \dfrac{4}{3}Rd\right)\{\theta'(\eta)\}^2 + B_r\left[\dfrac{4}{\mathrm{Re}_r}\{F(\eta)\}^2 + \left(\{F'(\eta)\}^2 + \{H'(\eta)\}^2\right)\right] + B_r K\left(\{F(\eta)\}^2 + \{H(\eta)\}^2\right)}. \quad \textbf{(24)}$$

SOLUTION PROCEDURE

In this part, the ODEs (14-16) with BCs (17) have been talked about involving BVP4C procedures in MATLAB, and graphical outcomes are analyzed. The graphs are plotted to look at the varieties of pertinent parameters. A flow outline of the resolution route is indicated in Fig. (**2**).

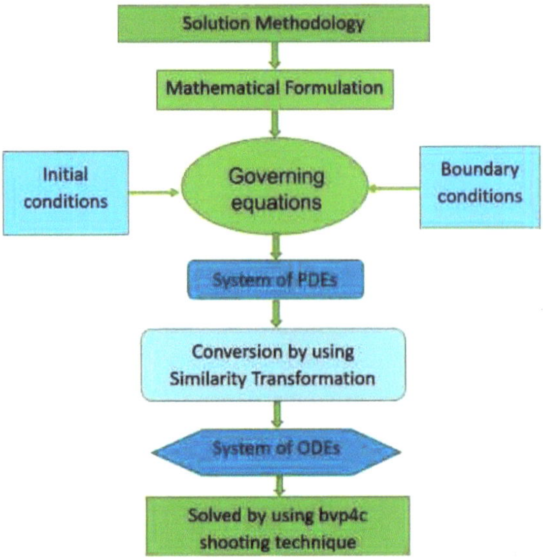

Fig. (2). The flow diagram of the solution process.

Initiating the new variables

$$y_1 = F, y_2 = F', y_3 = F'', y_4 = H, y_5 = H', y_6 = \theta, y_7 = \theta', \tag{25}$$

The 'dydx' function is given in the matrix form by the higher order derivative of every equation composed of lower orders as follows,

$$dydx = \begin{pmatrix} y_2 \\ y_3 \\ \dfrac{\lambda y_1 - y_1^2 + y_4 y_2 + (1+F_r) y_2^2 + \varepsilon_4 M y_1}{\frac{\varepsilon_1 \varepsilon_2}{2}} \\ y_5 \\ 2y_1 \\ y_7 \\ -\dfrac{-4\dfrac{Rd}{\Pr}(1+(\theta_w - 1)y_6)^2(\theta_w - 1)y_7^2 + \Pr \varepsilon_3 y_1 y_7 - 2\Pr R_T\left(y_1^2 y_6 + y_2 y_4 y_7\right) + Q_T y_6 \exp(-n\eta) + \dfrac{Br}{\varepsilon_1 \varepsilon_2}\left(y_2^2 + y_5^2\right)}{\dfrac{k_{hnf}}{k_f} + \dfrac{4}{3}\varepsilon_4 Rd\left(1+(\theta_w - 1)y_6\right)^3} \end{pmatrix}. \tag{26}$$

The initial guess is stated as:

$$\text{guess} = \left[\ -2\text{Ma}\varepsilon_1 \quad s_3 \ 1 \ s_5 \quad \right], \tag{27}$$

with unknown guesses s_3, s_5.

Utilizing the functions 'solinit = bvpinit (linspace (0, 1, 10), guess)' with the guess in Eq (27). The principal error and comparative errors are taken as $|\text{E}| < 10^{10}$.

RESULT AND ANALYSIS

As shown in Figs. (**3-19**), we have associated the general and hybrid nanofluid for each graph to demonstrate the impacts of the numerous governing factors on the system. Additionally, we have talked about the outcomes of skin friction and the physical parameters' rate of heat transmission, as shown in Tables **2** and **3**. The general values of the parameterare as follows $M = 2$, $K = 0.5$, $Rd = 1.0$, $Ma = 0.3$, $R_T = 0.5$, $Fr = 0.5$, $Q_T = 0.2$, $\Pr = 7$, $\theta_w = 0.9$, $K = 0.2$, $Br = 0.3$.

Velocity Profile

The flow velocity is opposed by the larger consequence of the magnetic variable (M) as shown in Fig. (**3**). The Lorentz force may oppose the fluid's motion and lower velocity, which is caused by the fluid's magnetization. Fig. (**4**). demonstrates how the non-uniform inertia parameter (Fr) affects the velocity distribution. The inertia coefficient is directly related to medium porosity and drag coefficient. As compared to 0.65, which has less of an influence, an F_r of 0.70 significantly improves the momentum border boundary layer. Ullah [7] made a comparable observation. The effect of velocity for various Marangoni parameter (Ma) results is shown in Fig. (**5**). An introduction of Marangoni convection supported intermolecular forces between the fluid molecules basically due to the presence of a gradient of the surface tension. This convective force restricts the movement of fluid particles and hence a drop was observed in the fluid flow velocity. Fig. (**6**) clarifies how the velocity profile decreases as the porosity variable (λ) increases in value. The fluid's velocity profile drops and decreases mentum boundary layer thickness due to the bigger value of the porosity parameter λ, which gives the same volume of fluid and more area to flow. The Prandtl number's velocity was investigated in Fig. (**7**). This is the ratio of momentum diffusivity (kinematic viscosity) to heat diffusivity.The relative thickness of the momentum boundary layers is controlled by the Prandtl number. With a reduction in velocity profile, the Prandtl number rises.

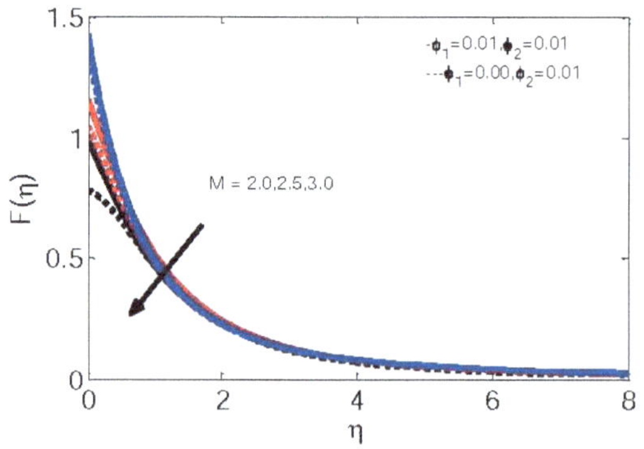

Fig. (3). Variation in velocity profile for different *M*.

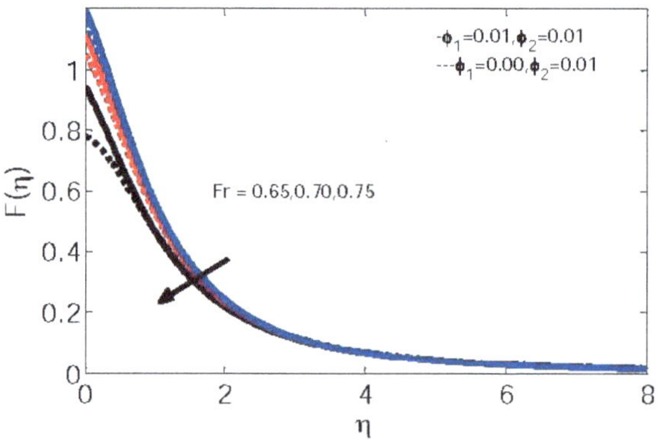

Fig. (4). Variation in velocity profile for different *Fr*.

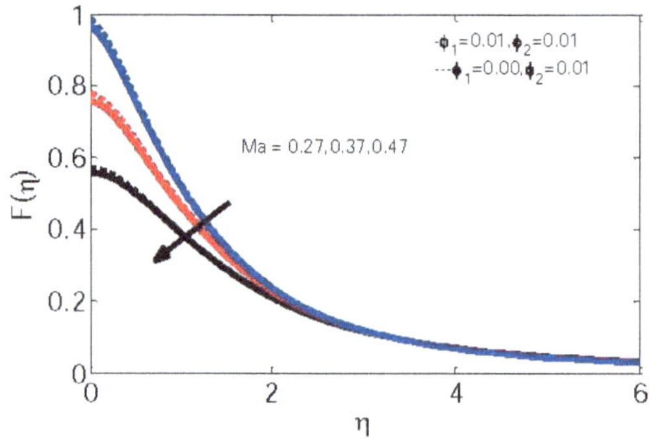

Fig. (5). Variation in velocity profile for different *M a*.

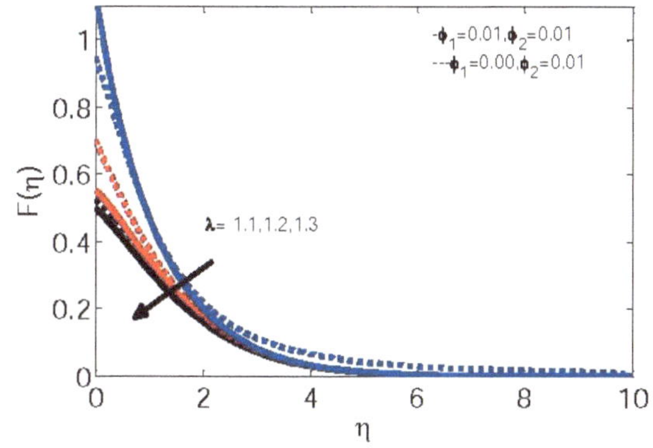

Fig. (6). Variation in velocity profile for different λ .

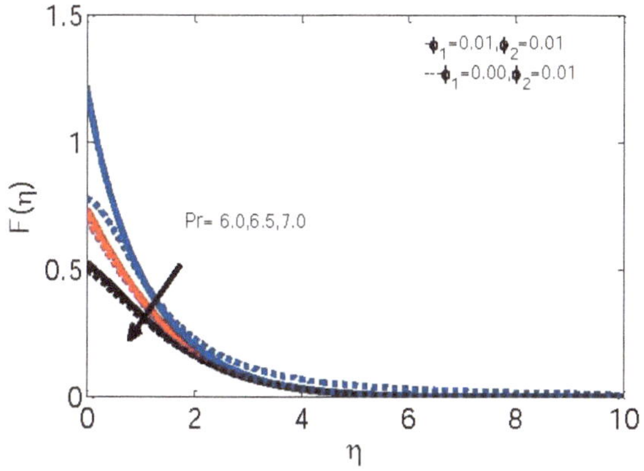

Fig. (7). Variation in velocity profile for different $\Pr$.

Temperature Profile

As the Marangoni parameter (Ma) was increased, the temperature profile decreased, as shown in Fig. (**8**). The movement of the molecules of the liquid became slower with an increase in Ma, which is possibly the reason of the drop in the thermal boundary layer thickness. For a range of non-uniform inertia coefficient (F_r) values, the temperature features are shown in Fig. (**9**). The thermal boundary layer width notices a drop in growth F_r and it may be due to the presence of inertia force in the system which controls the temperature of the fluid. Fig. (**10**) demonstrates that for a rise in magnetic parameter (M), the temperature profile also increased. Lorentz force exerted on the fluid by the magnetic parameter resulted in the suppression of the thermal boundary layer's width. Fig. (**11**) displays that the temperature increased, for higher values of the porosity variable which is quite opposite to the nature of the inertia term. However, the changes are not very significant. (Fig. **12**). Reveals the temperature profile for the Prandtl number. Prandtl number controls the heat diffusivity and hence with enhanced outcomes of the Prandtl number, the temperature outline is boosted. The temperature profile demonstrated in Fig. (**13**) improves when the thermal relaxation parameter R_T increases. Thermal relaxation is related to Cattaneo-Christov heat flux, which introduces an extra heat flux in the system, which consequently increases the temperature of the fluid. It is clear that an upsurge in the radiation parameter initiates the temperature profiles to rise see (Fig. **14**). Physically heat is created within the liquid $\gamma - ALOOH\,\&\,AA7072$ due to the presence of $\sigma^* T_\infty^{\ 3}$ which includes a further heat source in the system and supports the growth of the temperature in the fluid. The process of turning one form of energy into thermal energy within the layer is known as the creation of heat. Fig. (**15**). illustrates how temperature will rise with increasing Q. The heat generation generates heat in the system and controls the thermal stability of the system. As a consequence of the heat source procedure, we have observed that the hybrid nano liquids have better thermal conductivity in every instance, leading to the actual production of heat within the liquid. As such, the use of two particles contributes to a more stable system temperature management than does the presence of one particle.

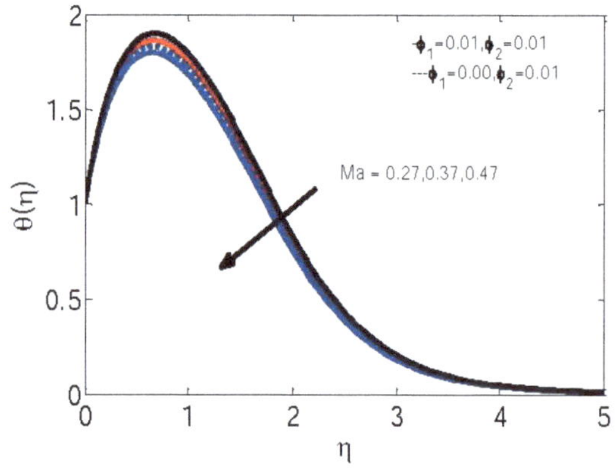

Fig. (8). Variation in temperature profile for different Ma.

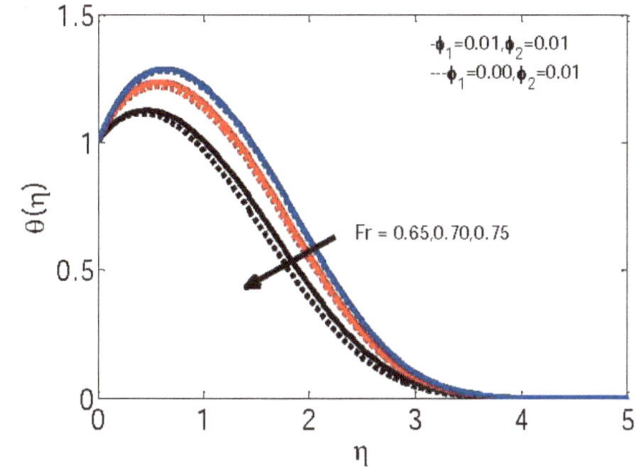

Fig. (9). Variation in temperature profile for different *Fr*.

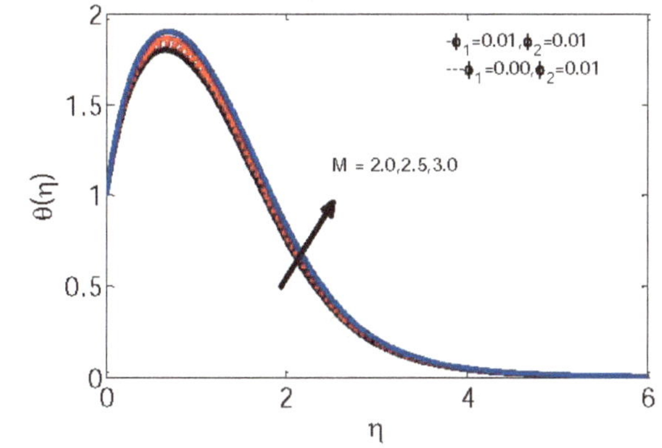

Fig. (10). Variation in temperature profile for different *M*.

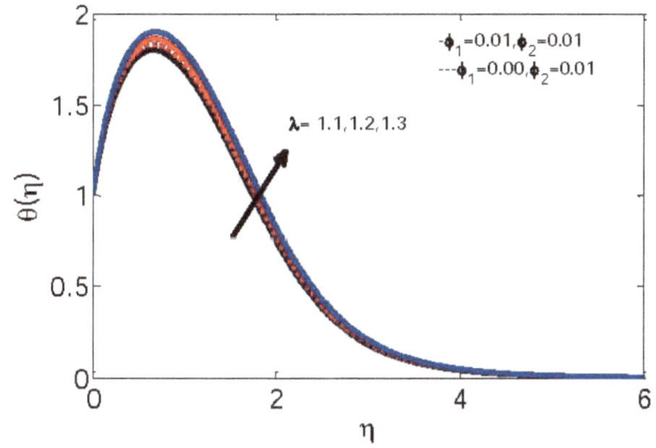

Fig. (11). Variation in temperature profile for different λ .

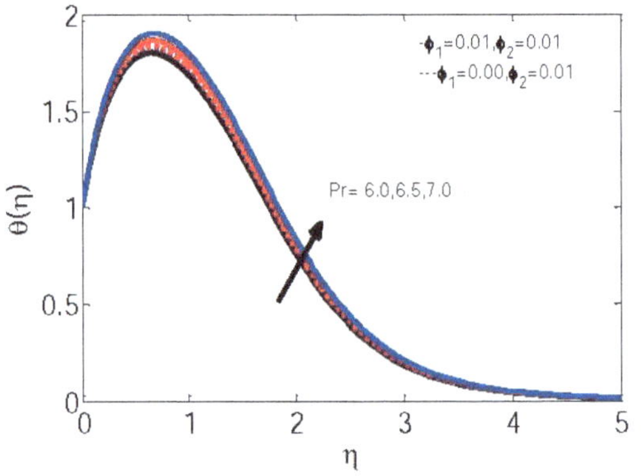

Fig. (12). Variation in temperature profile for different $\Pr$.

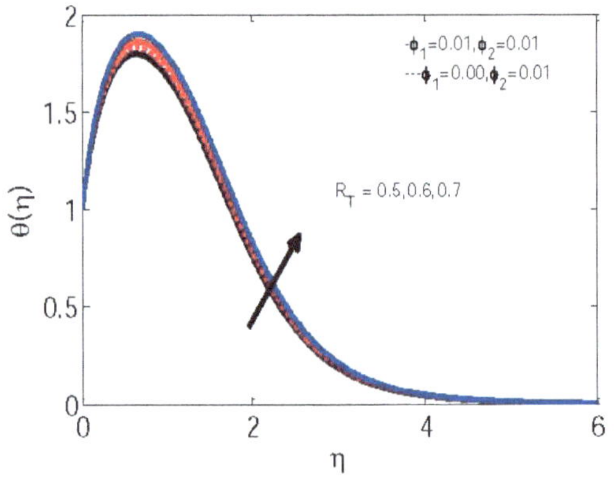

Fig. (13). Variation in temperature profile for different R_T .

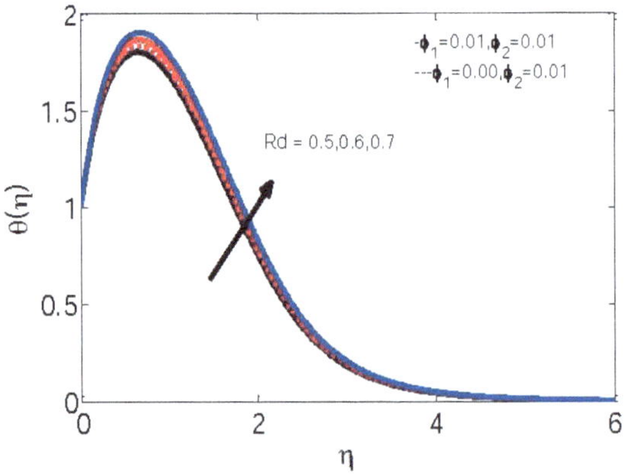

Fig. (14). Variation in temperature profile for different *Rd.*

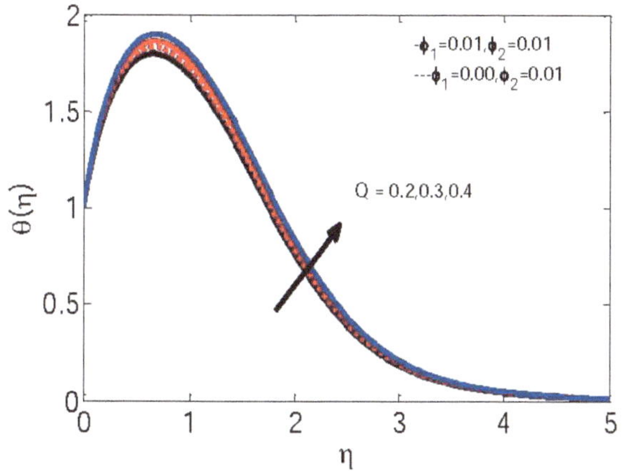

Fig. (15). Variation in temperature profile for different Q.

Skin Friction and Nusselt Number Profile

The skin friction profile (C_{fr}) for M and lambda is shown in Fig. (**16**). As M and λ values increased, the skin friction profile reduced. It is because of Lorentz force which is created by the magnetic parameter, which considerably lowers skin friction by reducing internal friction between fluid molecules. The profile of Nusselt numbers that varies with Rd and θ_w is shown in Fig. (**17**). The Nusselt number profile (Nu_r) rises with the increasing values of Rd and θ_w. Additionally, it was shown that compared to regular nanofluid, hybrid nanofluid supports the rate of heat transfer. The profile of Nusselt numbers that vary with Q_T and R_T is shown in Fig. (**18**). The Nusselt number profile improves with increasing values of Q_T and R_T. Additionally, Tables **1-2** contain the physical values of local Nusselt and skin friction for various parameters. We have assessed both the nanofluid as well as hybrid nanofluid. Table **1** displays the properties of nanofluid as well as its base fluid. Hence several values of water, Boehmite alumina $(\gamma-ALOOH)$, and Aluminium alloys $(AA7072)$ are considered. So, we applied the thermophysical properties to display the comparison of both nanofluid and hybrid nanofluid. Skin friction is shown in Table **2** for a range of factors that change with the hybrid and nanofluid. With replication, it is initiated that with an improvement in M , λ , F_r , Pr , θ_w, K , R_T , Q_T, and Ma, the skin friction becomes superior for general and hybrid nanofluid and is reduced with respect to the radiation parameter Rd . Further, the transfer rate of heat is increased with larger outcomes of M , λ , θ_w, K , R_T , Q_T and Ma and the same way, it is decreased for F_r and Pr .

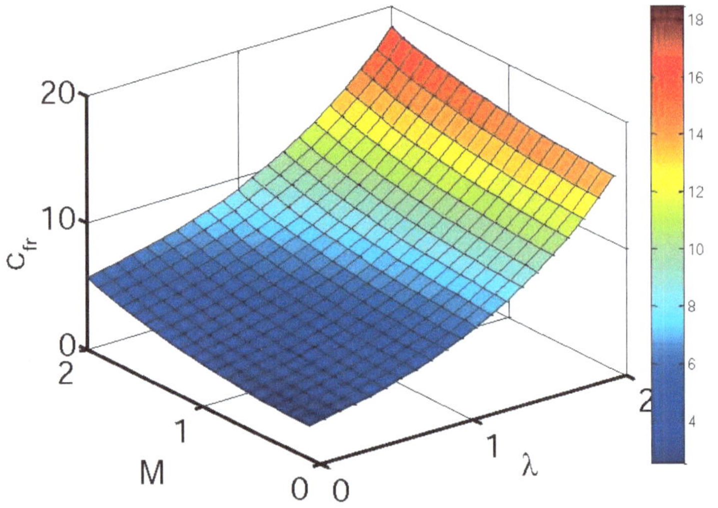

Fig. (16). Variation in Skin friction profile for different M, λ .

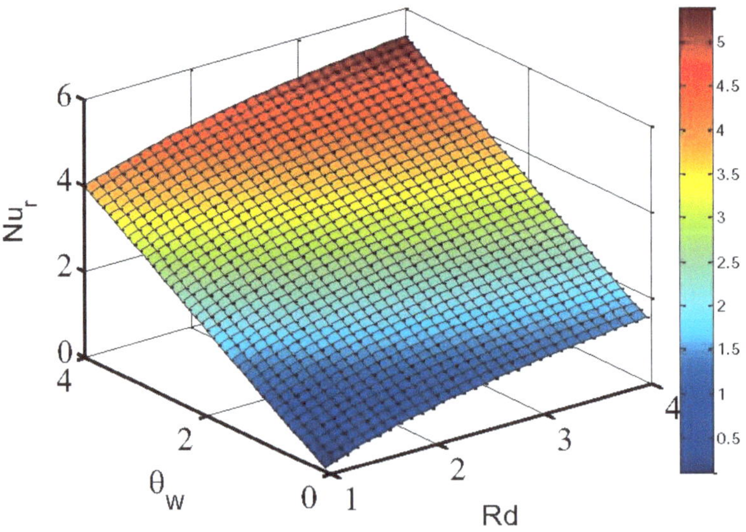

Fig. (17). Variation in Nusselt number profile for different Rd, θ_w .

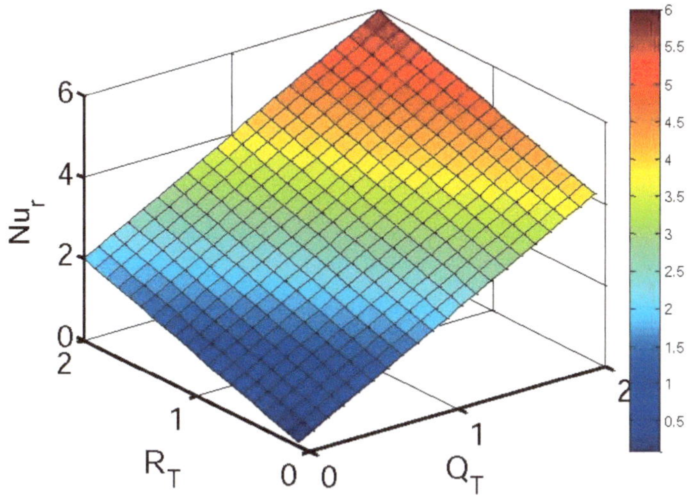

Fig. (18). Variation in Nusselt number profile for different Q_T, R_T .

Entropy Generation

A material with a high permeability allows fluids to pass through it more readily than one with a low permeability. The fluid velocity drops as an irreversible process as a consequence of permeability entropy, as shown in Figs. (**19** and **20**) demonstrates how the fast rate of heat transfer causes the Brinkman number to decrease as the entropy declines. With the permeability variable and Brinkman number, the Bejan number rises in Figs. (**21** and **22**), respectively. Hence the intermolecular tensions between the fluid molecules are enhanced.

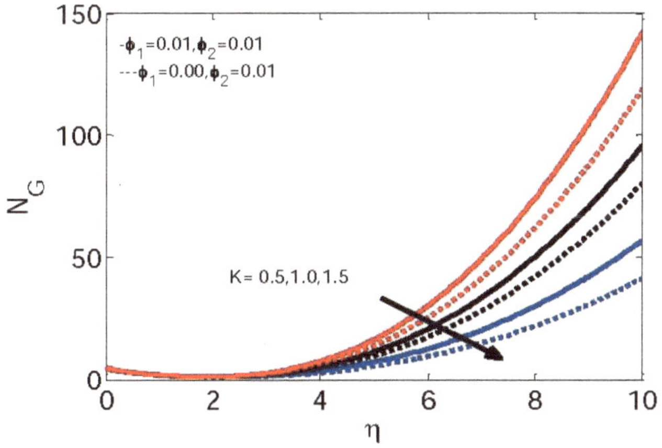

Fig. (19). Variation in entropy for different K .

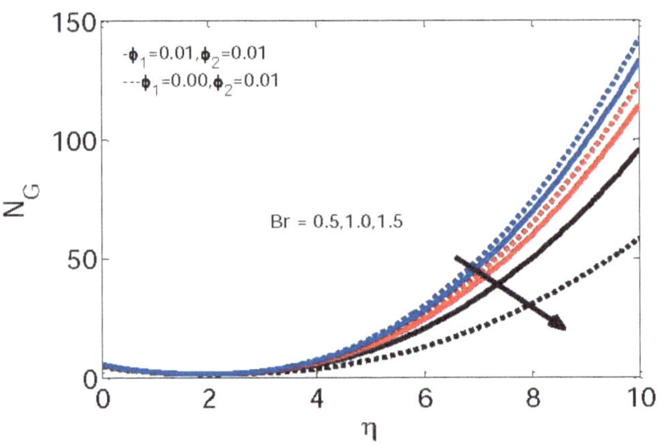

Fig. (20). Variation in entropy for different Br .

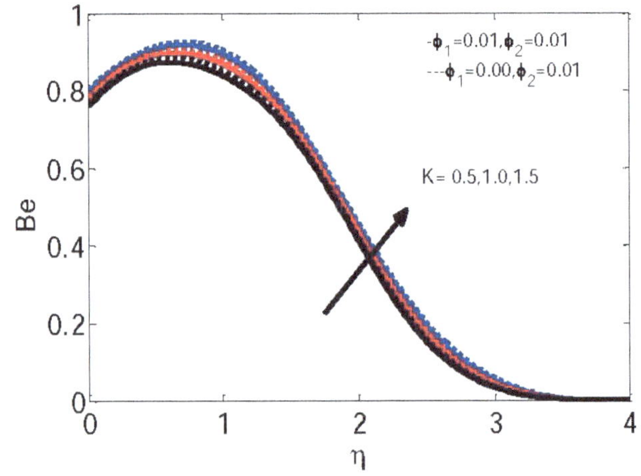

Fig. (21). Variation in Bejan number for different K .

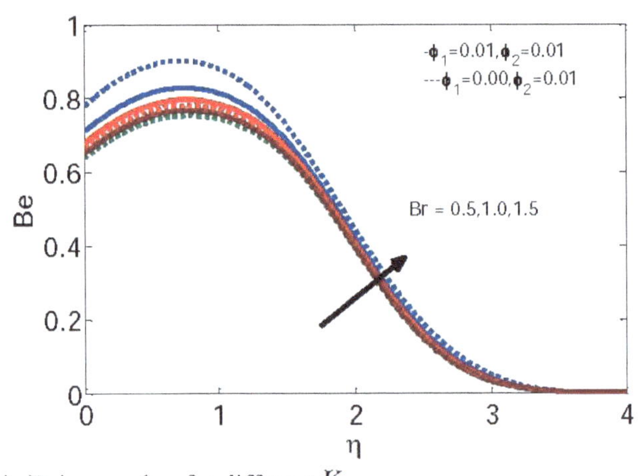

Fig. (22). Variation in Bejan number for different K .

Validation

We have equated our results with Ullah [7] for $-\theta'(0)$ about some conditions as detected in Table **3**, and a very valuable qualified value has been observed.

Table 1. Thermophysical characteristics of the base fluid and nano materials.

-	$\rho(Kg/m^3)$	$c_p(J/KgK)$	$k(W/mK)$	$\beta(K^{-1})$	$\sigma(\Omega^{-1}m^{-1})$
Water	1063.8	3630	0.387	5.8×10^{-4}	9.75×10^{-4}
$(\gamma-ALOOH)$	3050	618.3	30	21×10^{-5}	16.77×10^{6}
$(AA7072)$	2720	893	222	2.84×10^{-5}	34.83×10^{6}

Table 2. Skin friction and Heat transfer for various parameters with an assessment of nanofluid and hybrid nanofluid.

Parameters		$\varphi_1=0,\varphi_1=0.01,$		$\varphi_1=0.01,\varphi_1=0.01$	
		Skin friction	Heat transfer	Skin friction	Heat transfer
λ	1.1	0.05758030	0.02863705	0.05848549	0.02507660
	1.2	0.07781819	0.06822205	0.07090314	0.05263467
	1.3	0.07881819	0.06922205	0.09475302	0.08341458
M	2.0	3.57937108	0.19776191	3.73156208	0.23436312
	2.5	6.51203161	0.58732735	7.74989124	0.52937483
	3.0	7.85709073	0.46122917	11.9993747	0.53763634
Fr	0.65	0.07608781	0.06783223	0.07635542	0.06305115
	0.70	0.08161756	0.06909168	0.08068037	0.06381523

(Table 2) cont.....

	0.75	0.08681738	0.07015659	0.09068036	0.07381587
Pr	6.0	0.09589295	0.03786981	0.09524713	0.03483646
	6.5	0.07693579	0.06802018	0.07716085	0.06318136
	7.0	0.07427925	0.09710379	0.07464243	0.09053988
Rd	0.5	0.06778541	0.05893382	0.07716085	0.06318136
	1.0	0.02484039	0.01158233	0.02881049	0.01324167
	1.5	0.01606242	0.00052474	0.01841832	0.00088638
θ_w	0.3	0.15972846	0.14771629	0.06847691	0.05489008
	0.5	0.14144304	0.13142303	0.03974951	0.02456301
	0.7	0.10846993	0.10040106	0.10846993	0.10040106
K	0.5	0.06533511	0.05789353	0.06847691	0.05489008
	1.0	0.04287288	0.04368864	0.04347655	0.04361196
	1.5	0.06993259	0.05995381	0.07038783	0.02722856
R_T	0.5	0.07703462	0.06620552	0.07314603	0.05768360
	0.6	0.10042738	0.09007775	0.10385469	0.09584248
	0.7	0.12086173	0.10998463	0.12142060	0.10434838
Q	0.2	0.01809724	0.00032277	0.07634441	0.06083488
	0.3	0.06887907	0.05770775	0.07090648	0.05546498
	0.4	0.07887907	0.06770775	0.36282840	0.28191146
Ma	0.27	0.07951665	0.02764617	0.06925211	0.05381969
	0.37	0.44986791	0.00579496	0.33679940	0.00204425
	0.47	2.06041547	0.06329565	1.70173297	0.05332340

Table 3. Comparison of λ with $-\theta(0)$ Ullah [7] for $F_r = M = Rd = R_T = Q_T = Br = 0$, $\Pr = 7$ and $\varepsilon_1 = 1$.

λ	Ullah [7]	Present
0.1	2.264607	2.2646071
0.3	1.997294	1.9973242
0.4	1.665655	1.6657624

CONCLUSION

The Boehmite alumina and aluminium alloy are used in the research for thermal Marangoni convection with steady Darcy-Forchheimer flow with gasoline oil. By utilizing boundary conditions, the preceding PDEs are converted into a set of ODEs on the implication of similarity transformations and then resolved numerically. Generally, in the process of irreversibility, we implement the entropy generation and Bejan number to display the high heat transfer rate of the flow. The thermos-physical features are detected to establish the impact of nanoparticles in the Marangoni convection flow. Every graph demonstrates a brief assessment of nanofluid and hybrid nanofluid with various parameters. The major concepts of the problem are stated as;

- Velocity profiles diminished with $Fr, M, Ma, \Pr$, λ.
- The temperature profiles intensified with $Pr, M, \lambda, Rd, Q, R_T, Ma$ and diminish with *Fr*.
- The skin friction increases with rising rates of M, λ, Fr, Pr, θ w, K, R_T, Q, and Ma and dimin ishes with higher outcomes of Pr, *Fr, Rd*.
- The transfer rate of heat increases with larger outcomes of *M, Fr, Pr, K, Q* and decreases with larger outcomes of *λ, Rd, θ w,* R_T, and *Ma*.

These conclusions may assist in recognizing various industrial occurrences in semiconductor processing, silicon wafers, atomic reactor thin-film stretching, *etc*.

REFERENCES

[1] S.U.S. Choi, Z.G. Zhang, W. Yu, F.E. Lockwood, and E.A. Grulke, "Anomalous thermal conductivity enhancement in nanotube suspensions", *Appl. Phys. Lett.,* vol. 79, no. 14, pp. 2252-2254, 2001.

http://dx.doi.org/10.1063/1.1408272

[2] M. K. Nayak, and G. Mahanta, "Entropy analysis of a 3D nonlinear radiative hybrid nanofluid flow between two parallel stretching permeable sheets with slip velocities", *Int. J. Ambient energy,* vol. 43, no. 1, pp. 8710-8721, 2022.

[3] L. Zhang, M.M. Bhapti, E.E. Michal Lides, M. Marine, and R. Elahi, "Hybrid nanofluid flow towards an elastic surface with tantalum and nickel nanoparticles, under the influence of an induced magnetic field", *Eur. Phys. J. Spec. Top.,* pp. 1-13, 2021.

[4] S. Hosseinzadeh, and D.D. Ganji, "A novel approach for assessment of MHD mixed fluid around two parallel plates by consideration hybrid nanoparticles and shape factor", *Alex. Eng. J.,* vol. 61, no. 12, pp. 9779-9793, 2022.

http://dx.doi.org/10.1016/j.aej.2022.03.031

[5] M. M. Bhati, O. Beg, and S. I. Abdel Salam, *Nano material,* vol. 12, p. 1049, 2022.

[6] R. Md Kasmani, S. Sivasankaran, M. Bhuvaneswari, and A.K. Hussein, "Analytical and numerical study on convection of nanofluid past a moving wedge with Soret and Dufour effects", *Int. J. Numer. Methods Heat Fluid Flow,* vol. 27, no. 10, pp. 2333-2354, 2017.

http://dx.doi.org/10.1108/HFF-07-2016-0277

[7] I. Ullah, "Heat transfer enhancement in Marangoni convection and nonlinear radiative flow of gasoline oil conveying Boehmite alumina and aluminum alloy nanoparticles", *Int. Commun. Heat Mass Transf.,* vol. 132, no. 3, p. 105920, 2022.

http://dx.doi.org/10.1016/j.icheatmasstransfer.2022.105920

[8] B. Sehar, A. Waris, S.O. Gilani, U. Ansari, S. Mushtaq, N.B. Khan, M. Jameel, M.I. Khan, O.T. Bafakeeh, and E.S.M. Tag-ElDin, "The impact of laminations on the mechanical strength of carbon-fiber composites for prosthetic foot fabrication", *Crystals* , vol. 12, no. 10, p. 1429, 2022.

http://dx.doi.org/10.3390/cryst12101429

[9] M. Shahid, H.M.A. Javed, M.I. Ahmad, A.A. Qureshi, M.I. Khan, M.A. Alnuwaiser, A. Ahmed, M.A. Khan, E.S.M. Tag-ElDin, A. Shahid, and A. Rafique, "A brief assessment on recent developments in efficient electrocatalytic Nitrogen reduction with 2D non-metallic nanomaterials", *Nanomaterials* , vol. 12, no. 19, p. 3413, 2022.

http://dx.doi.org/10.3390/nano12193413 PMID: 36234541

[10] S. Suresh, K.P. Venkitaraj, P. Selvakumar, and M. Chandrasekar, "Effect of Al_2O_3–Cu/water hybrid nanofluid in heat transfer", *Exp. Therm. Fluid Sci.,* vol. 38, pp. 54-60, 2012.

http://dx.doi.org/10.1016/j.expthermflusci.2011.11.007

[11] S. Suresh, K.P. Venkitaraj, M.S. Hameed, and J. Sarangan, "Turbulent heat transfer and pressure drop characteristics of dilute water based Al_2O_3-Cu hybrid nanofluids", *J. Nanosci. Nanotechnol.,* vol. 14, no. 3, pp. 2563-2572, 2014.

http://dx.doi.org/10.1166/jnn.2014.8467 PMID: 24745264

[12] M. Baghbanzadeh, A. Rashidi, D. Rashtchian, R. Lotfi, and A. Amrollahi, "Synthesis of spherical silica/multiwall carbon nanotubes hybrid nanostructures and investigation of thermal conductivity of related nanofluids", *Thermochemical acta,* vol. 549, p. 87, 2012.

http://dx.doi.org/10.1016/j.tca.2012.09.006

[13] M. J. Nine, M. Batmunkh, J. H. Kim, H. S. Chung, and H. M. Jeong, "Investigation of Al_2O_3-MWCNTs hybrid dispersion in water and their thermal characterization", *J. nanoscience and nanotech.,* vol. 12, p. 4553, 2012.

[14] A. Abbasi, W. Farooq, E.S.M. Tag-ElDin, S.U. Khan, M.I. Khan, K. Guedri, S. Elattar, M. Waqas, and A.M. Galal, "Heat transport exploration for hybrid nanoparticle (Cu, Fe_3O_4)—Based blood flow *via* tapered complex wavy curved channel with slip features", *Micromachines* , vol. 13, no. 9, p. 1415, 2022.

http://dx.doi.org/10.3390/mi13091415 PMID: 36144038

[15] S. Nadeem, and N. Abbas, "On both MHD and slip effect in micropolar hybrid nanofluid past a circular cylinder under stagnation point region", *Can. J. Phys.,* vol. 97, no. 4, pp. 392-399, 2019.

http://dx.doi.org/10.1139/cjp-2018-0173

[16] N. Abbas, S. Saleem, S. Nadeem, A.A. Alderremy, and A.U. Khan, "On stagnation point flow of a micro polar nanofluid past a circular cylinder with velocity and thermal slip", *Results Phys.,* vol. 9, pp. 1224-1232, 2018.

http://dx.doi.org/10.1016/j.rinp.2018.04.017

[17] D. Mohanty, G. Mahanta, and S. Shaw, "Analysis of irreversibility for 3-D MHD convective Darcy–Forchheimer Casson hybrid nanofluid flow due to a rotating disk with Cattaneo–Christov heat flux, Joule heating, and nonlinear thermal radiation", *Numer. Heat Transf. B,* vol. 84, no. 2, pp. 115-142, 2023.

http://dx.doi.org/10.1080/10407790.2023.2189644

[18] S. Nadeem, and N. Abbas, "Effects of MHD on modified nanofluid model with variable viscosity in a porous medium", *Nanof. Flow in Por. Med.,* vol. 7, 2019.

[19] S. Nadeem, M.Y. Malik, and N. Abbas, "Heat transfer of three-dimensional micropolar fluid on a Riga plate", *Can. J. Phys.,* 2019.

[20] D. Mohanty, G. Mahanta, S. Shaw, and M. Das, "Thermosolutal Marangoni stagnation point GO–MoS2/water hybrid nanofluid over a stretching sheet with the inclined magnetic field", *Int. J. Mod. Phys. B,* p. 2450024, 2023.

[21] N. A. M. Malik, and S. Nadeem, "Study of three-dimensional stagnation point flow of hybrid nanofluid over an isotropic slip surface", 2019.

[22] F.R. Mazarron, C.J. Porras-Prieto, J.L. Garcia, and R.M. Benavente, "Feasibility of active solar water heating systems with evacuated tube collector at different operational water temperatures", *Energy Convers. Manag,* vol. 113, 2016.

[23] A. Shafieian, M. Khiadani, and A. Nosrati, "Strategies to improve the thermal performance of heat pipe solar collectors in solar systems: A review", *Energy Convers. Manage.,* vol. 183, pp. 307-331, 2019.

http://dx.doi.org/10.1016/j.enconman.2018.12.115

[24] K.M. Pandey, and R. Chaurasiya, "A review on analysis and development of solar flat plate collector", *Renew. Sustain. Energy Rev.,* vol. 67, pp. 641-650, 2017.

http://dx.doi.org/10.1016/j.rser.2016.09.078

[25] X. Nie, L. Zhao, S. Deng, and X. Lin, "Experimental study on thermal performance of U-type evacuated glass tubular solar collector with low inlet temperature", *Sol. Energy,* vol. 150, 2017.

[26] V.K. Jebasingh, and G.M.J. Herbert, "A review of solar parabolic trough collector", *Energy Rev,* vol. 54, 2016.

[27] S.H. Farjana, N. Huda, M.A.P. Mahmud, and R. Saidur, "Solar industrial process heating systems in operation Current SHIP plants and future prospects in Australia", *Renew. Sustain. Energy Rev.,* vol. 91, pp. 409-419, 2018.

http://dx.doi.org/10.1016/j.rser.2018.03.105

[28] D. Cocco, M. Petrollese, and V. Tola, "Exergy analysis of concentrating solar systems for heat and power production", *Energy, vol.* 130, 2017.

[29] M. Mehrpooya, B. Ghorbani, and S.S. Hosseini, "Thermodynamic and economic evaluation of a novel concentrated solar power system integrated with absorption refrigeration and desalination cycles", *Energy Convers. Manag,* vol. 175, 2018.

[30] M. A. Serag-Eldin, "Thermal design of a roof-mounted CLFR collection system for a desert absorption chiller", *Int. J. Sustain. Energy,* vol. 33, 2012.

[31] E. Przenzak, M. Szubel, and M. Filipowicz, "The numerical model of the high temperature receiver for concentrated solar radiation", *Energy Convers. Manag,* vol. 125, 2016.

[32] M.S. Shadloo, A. Hadjadj, and F. Hussain, "Statistical behavior of supersonic turbulent boundary layers with heat transfer at M $\infty = 2$", *Int. J. Heat Fluid Flow,* vol. 53, pp. 113-134, 2015.

http://dx.doi.org/10.1016/j.ijheatfluidflow.2015.02.004

[33] A. Rahmat, N. Tofighi, M.S. Shadloo, and M. Yildiz, "Numerical simulation of wall bounded and electrically excited Rayleigh–Taylor instability using incompressible smoothed particle hydrodynamics", *Colloids Surf. A Physicochem. Eng. Asp.,* vol. 460, pp. 60-70, 2014.

http://dx.doi.org/10.1016/j.colsurfa.2014.02.044

[34] A. Zainali, N. Tofighi, M.S. Shadloo, and M. Yildiz, "Numerical investigation of Newtonian and non-Newtonian multiphase flows using ISPH method", *Comput. Methods Appl. Mech. Eng.,* vol. 254, pp. 99-113, 2013.

http://dx.doi.org/10.1016/j.cma.2012.10.005

[35] L.S. Sundar, and K.V. Sharma, "Turbulent heat transfer and friction factor of Al_2O_3 Nanofluid in circular tube with twisted tape inserts", *Int. J. Heat Mass Transf.,* vol. 53, no. 7-8, pp. 1409-1416, 2010.

http://dx.doi.org/10.1016/j.ijheatmasstransfer.2009.12.016

[36] M.F. Ahmed, A. Zaib, F. Ali, O.T. Bafakeeh, E.S.M. Tag-ElDin, K. Guedri, S. Elattar, and M.I. Khan, "Numerical computation for gyrotactic microorganisms in MHD radiative Eyring–Powell nanomaterial flow by a static/moving wedge with Darcy–Forchheimer relation", *Micromachines ,* vol. 13, no. 10, p. 1768, 2022.

http://dx.doi.org/10.3390/mi13101768 PMID: 36296121

[37] M. Hejazian, M.K. Moraveji, and A. Beheshti, "Comparative numerical investigation on TiO2/water nanofluid turbulent flow by implementation of single phase and two-phase approaches", *Numer. Heat Transf. A,* vol. 66, no. 3, pp. 330-348, 2014.

http://dx.doi.org/10.1080/10407782.2013.873271

[38] A. Bejan, CRC Press Boca Raton Florida, 1966.

[39] V. Bianco, O. Manca, and S. Nardini, "Entropy generation analysis of turbulent convection flow of Al_2O_3–water nanofluid in a circular tube subjected to constant wall heat flux", *Energy Convers. Manage.,* vol. 77, pp. 306-314, 2014.

http://dx.doi.org/10.1016/j.enconman.2013.09.049

[40] D. Mohanty, N. Sethy, G. Mahanta, and S. Shaw, "Impact of the interfacial nanolayer on Marangoni convective Darcy-Forchheimer hybrid nanofluid flow over an infinite porous disk with Cattaneo-Christov heat flux", *Therm. Sci. Eng. Prog.,* vol. 41, p. 101854, 2023.

http://dx.doi.org/10.1016/j.tsep.2023.101854

[41] M.M. Rashidi, N. Kavyani, and S. Abelman, "Investigation of entropy generation in MHD and slip flow over a rotating porous disk with variable properties", *Int. J. Heat Mass Transf.,* vol. 70, pp. 892-917, 2014.

http://dx.doi.org/10.1016/j.ijheatmasstransfer.2013.11.058

[42] O. Mahian, A. Kianifar, C. Kleinstreuer, M.A. Al-Nimr, I. Pop, A.Z. Sahin, and S. Wongwises, "A review of entropy generation in nanofluid flow", *Int. J. Heat Mass Transf.,* vol. 65, pp. 514-532, 2013.

http://dx.doi.org/10.1016/j.ijheatmasstransfer.2013.06.010

[43] M. Goodarzi, M.R. Safaei, H.F. Oztop, A. Karimipour, E. Sadeghinezhad, M. Dahari, S.N. Kazi, and N. Jomhari, "Numerical study of entropy generation due to coupled laminar and turbulent mixed convection and thermal radiation in an enclosure filled with a semitransparent medium", *Scient.World.J.,* vol. 2014, pp. 1-8, 2014.

http://dx.doi.org/10.1155/2014/761745 PMID: 24778601

[44] I. Siddique, S. Abdal, I.S.U. Din, J. Awrejcewicz, W. Pawkokski, and S. Hussain, "Significance of concentration-dependent viscosity on the dynamics of tangent hyperbolic nanofluid subject to motile microorganisms over a non-linear stretching surface", *Sci. Rep.,* vol. 12, no. 1, p. 12765, 2022.

http://dx.doi.org/10.1038/s41598-022-16601-9 PMID: 35896639

[45] B. Ali, N.A. Ahammad, A.U. Awan, A.S. Oke, E.M. Tag-ElDin, F.A. Shah, and S. Majeed, "The dynamics of water-based nanofluid subject to the nanoparticle's radius with a siwnificant magnetic field: The case of rotating micropolar fluid", *Sustainability,* vol. 14, no. 17, p. 10474, 2022.

http://dx.doi.org/10.3390/su141710474

CHAPTER 2

Three-Dimensional Numerical Computational Model for an Unsteady Hybrid Nanofluids Past Stretching MHD Rotating Sheet

Salma Ahmedai[1, 2*], Precious Sibanda[1], Sicelo Goqo[1] and Uthman Rufai[1]

[1]*School of Mathematics, Statistics and Computer Science, University of KwaZulu-Natal, Pietermaritzburg, South Africa*

[2]*Faculty of Mathematical Sciences and Statistics, Al Neelain University, Khartoum, Sudan*

Abstract: This paper presents a numerical approach to solving unsteady hybrid nanofluid flow problems using the overlapping grid multi-domain bivariate spectral simple iteration method (OMD-BSSIM). The method utilizes the Chebyshev spectral collocation approach to approximate the derivatives in the overlapping grid of space and non-overlapping grid of time, which allows for handling the two domains (multi-domain) problem. In this paper, we propose using the overlapping bivariate spectral method in combination with the simple iteration method instead of the commonly used quasi-linearization, relaxation, or local linearization schemes. The governing equations are transformed into a system of nonlinear partial differential equations using similarity transformations. The OMD-BSSIM is applied to investigate the heat transfer rate for MHD, unsteady $GNP - Fe_3O_4/H_2O$, and $TiO_2 - Fe_3O_4/H_2O$ hybrid nanofluids flow, with projection angles ranging from $0°$to$90°$, representing the influence of different magnetic fields. The numerical solution is obtained using OMD-BSSIM implemented in MATLAB. We use R visualization techniques and graphs to analyze the relationship between the skin friction coefficients, local Nusselt number, and Sherwood number of the hybrid nanofluids and some parameters. The results show that the increase in the volume fraction of GNP nanoparticle has a greater effect on the temperature profile than TiO_2 nanoparticle. Additionally, notable positive relationships were observed for the rotation parameter, while the stretching parameter had a negative impact on certain outcome measures.

Keywords: Hybrid nanofluid, Overlapping chebyshev spectral collocation, Rotating surface, Simple iteration method.

INTRODUCTION

In 1961, Sakiadis [1, 2] conducted studies on boundary layer flow for continuously moving surfaces, which included flat and cylindrical surfaces. His research

*Corresponding author Salma Ahmedai:** School of Mathematics, Statistics and Computer Science, University of KwaZulu-Natal, Pietermaritzburg, South Africa; Faculty of Mathematical Sciences and Statistics, Al Neelain University, Khartoum, Sudan; E-mail: sahmedai@neelain.edu.sd

Sabyasachi Mondal (Ed.)

introduced a new class of boundary layer problems with solutions that have a wide range of applications. The investigation of boundary layer flow has since become an important area of research in fluid mechanics, with numerous studies exploring the behaviour of boundary layers and their impact on heat transfer. Furthermore, Sakiadis's focus is on both laminar and turbulent flows contributed to the expansion of the field. In 1970, Crane [3] furthered Sakiadis's research by considering a stretching surface instead of a rigid one. Subsequently, multiple authors investigated these characteristics for different types of fluids, each utilizing a unique set of fluid flow properties and working with specific objectives, parameters, and assumptions.

Numerous studies have investigated stretching boundary layer flows [4, 5]. Recently, the use of nanofluids and hybrid nanofluids has gained significant popularity in contemporary industrial applications. The concept of nanofluids was introduced by Choi and Eastman in their pioneering work in 1995 [6]. A nanofluid is composed of a base fluid mixed with a single nanoparticle (NP) [7, 8]. Conversely, a hybrid nanofluid contains one or more additional NPs [9, 10]. Since 1995, these fluids have found a diverse range of industrial and domestic applications. Hybrid nanofluids, for instance, have significant applications in transformer cooling, building heating, and biomedical drug reduction, among others. Moreover, literature has demonstrated that the addition of nanoparticles into a liquid enhances its electrical, thermal, and acoustical conductivity [11]. Nanoparticles are materials characterized by their extremely small size, usually ranging from 1 to 100 nm [12].

In recent years, there has been significant interest in the development of materials that exhibit superconductivity at room temperature. This research area has garnered the attention of several scientists, who have been working on this intriguing field. Empirical studies have demonstrated that transition-metal oxides have promising thermoelectric properties. The development of such materials would enable the creation of cooler electronic components and significantly enhance the efficiency of the electricity grid. Through experimental investigations, Ma *et al.* [13] have made attempts to improve the hydrophobicity of graphene (GNP) nanoplatelets by fictionalization, without compromising their good thermal properties. This development has led to the creation of novel hybrid nanofluids that enhance heat transfer. Moreover, it effectively prevents the separation of composite nanoparticles during the boiling process. Ottofuelling *et al.* [14] investigated the colloidal stability of titanium dioxide (TiO_2) nanoparticles in different water environments and identified several factors that affect their aggregation behaviour. Their study demonstrates that the TiO_2 nanoparticle's stability is influenced by ionic strength, and the presence of natural organic matter and divalent ions. Meanwhile, Othman *et al.* [15] have shared an extraordinary finding that water vapor can induce superconductivity in iron arsenic-containing strontium at 25 *K*. Water appears to cause a kind of chemical or structural change that prompts the transition at a higher temperature.

Hannes Alfvén [16] is credited with formally introducing the field of magnetohydrodynamics (MHD). The MHD defines and studies the dynamics of electrically conducting fluids, mainly focusing on the control of heat transfer. Several authors have conducted numerical investigations on heat transfer properties of the MHD boundary layer flow. For example, Raza *et al.* [17] examined a micro-polar and MHD flow past stretching and shrinking sheets in a porous media with thermal radiation and suction/injection effects. They concluded that the temperature was directly proportional to the thermal radiation parameter, magnetic parameter, and Darcy number. On the other hand, it was inversely proportional to the suction parameter. Furthermore, Khan and Nadeem [18] studied double-stratified rotating Maxwell nanofluid flowing over both linear and exponential stretching sheets. They reported that an increase in the rotation and or relaxation parameters decreased fluid velocity but increased the temperature and related boundary layer thickness.

In this study, we investigated numerically two types of hybrid nanofluids, $GNP - Fe_3O_4/H_2O$, and $TiO_2 - Fe_3O_4/H_2O$ in changing magnetic angles. We utilized the overlapping grid multi-domain bivariate spectral simple iteration method (OMD-BSSIM). The modified method is a combination of two methods, the simple iteration method (SIM) as a linearizing technique and the overlapping multi-domain spectral collocation discretization. The spatial domain is divided into overlapping sub-domains, whereas the temporal domain is divided into non-overlapping sub-domains. The overlapping grid procedure leads to dense matrix equations and was recently proposed by Mkhatshwa *et al.* [19]. Some studies have validated the use of overlapping grid procedures [20, 21] to solve the nonlinear partial differential equations (N-PDEs). The SIM technique was proposed by Motsa *et al.* [22], and it is based on fixed point iteration to develop iterative schemes. The SIM has demonstrated practicality in solving differential equations with applications in fluid mechanics [23].

MATHEMATICAL MODEL

The 3D, MHD, and unsteady flow of an electrically conductive, incompressible, rotating, stratified Fe_3O_4/H_2O hybrid nanofluids flow past a stretching sheet. The flow is defined at $z_3 \geq 0$. Fig. (**1**) shows the coordinate system and the physical model of the problem. The characteristic length scale is taken to be the plate's length L as it represents the dimension in the direction of the imposed flow of the fluid. Further, Ω_0 and B_0 are the rotation and magnetic coefficients respectively. In this study, a constant angular velocity $\Omega_0$is maintained as the hybrid nanofluid flow undergoes rotation about the vertical axis. In addition, both T_0 and ambient temperature T_∞ are positive constants.

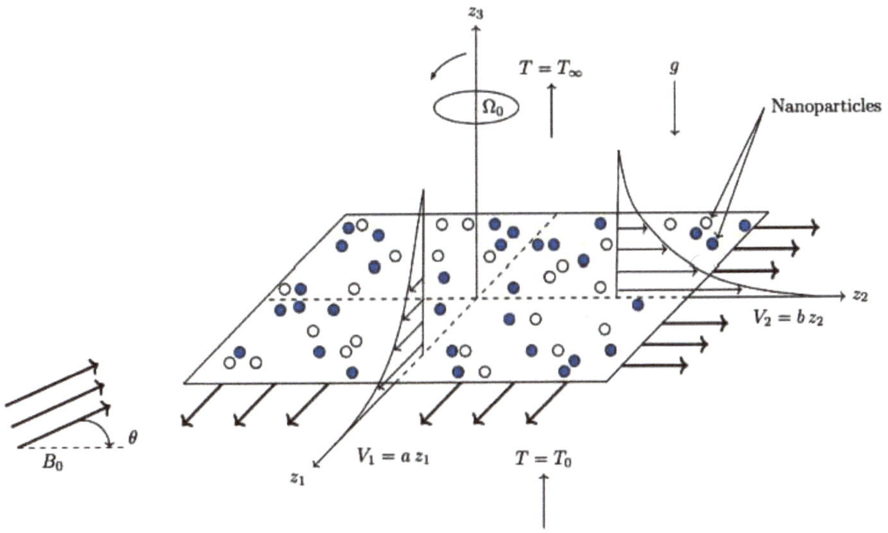

Fig. (1). Physical configuration and coordinate system.

The momentum, energy, concentration, and continuity equations for the system are provided based on the stated assumptions see [24, 25].

$$\frac{\partial v_1}{\partial t} + v_1 \frac{\partial v_1}{\partial z_1} + v_2 \frac{\partial v_1}{\partial z_2} + v_3 \frac{\partial v_1}{\partial z_3} - 2\,\Omega_0\, v_2 = \frac{\mu_{hnf}}{\rho_{hnf}} \frac{\partial^2 v_1}{\partial z_3^2} + \frac{\sigma_{hnf}}{\rho_{hnf}} B_0^2 \,(v_2 \sin\theta \cos\theta - v_1 \sin^2\theta) + g\beta_{hnf}\,(T - T_\infty), \tag{1}$$

$$\frac{\partial v_2}{\partial t} + v_1 \frac{\partial v_2}{\partial z_1} + v_2 \frac{\partial v_2}{\partial z_2} + v_3 \frac{\partial v_2}{\partial z_3} + 2\Omega_0 v_1 = \frac{\mu_{hnf}}{\rho_{hnf}} \frac{\partial^2 v_2}{\partial z_3^2} + \frac{\sigma_{hnf}}{\rho_{hnf}} B_0^2 \,(v_1 \sin\theta \cos\theta - v_2 \sin^2\theta), \tag{2}$$

$$\frac{\partial T}{\partial t} + v_1 \frac{\partial T}{\partial z_1} + v_2 \frac{\partial T}{\partial z_2} + v_3 \frac{\partial T}{\partial z_3} = \alpha_{hnf} \frac{\partial^2 T}{\partial z_3^2} + \frac{Q_0}{(\rho c_p)_{hnf}} (T - T_\infty), \tag{3}$$

$$\frac{\partial C}{\partial t} + v_1 \frac{\partial C}{\partial z_1} + v_2 \frac{\partial C}{\partial z_2} + v_3 \frac{\partial C}{\partial z_3} = D_{hnf} \frac{\partial^2 C}{\partial z_3^2}, \tag{4}$$

$$\frac{\partial v_1}{\partial z_1} + \frac{\partial v_2}{\partial z_2} + \frac{\partial v_3}{\partial z_3} = 0, \tag{5}$$

Here μ_{hnf}, ρ_{hnf}, κ_{hnf}, $(\rho\, c)_{\mathrm{hnf}}$, σ_{hnf} and D_{hnf} represent the dynamic viscosity, density, thermal conductivity, heat capacity, electrical conductivity, and diffusivity of the hybrid nanofluid respectively. In the current work, several assumptions are considered in the hybrid nanofluid model, which are listed below:

[1] Including the effects of heat generation and absorption.
[2] Spherical shape of the nanoparticles.
[3] The plate is extended equally in z_1 and z_2 directions.
[4] The projection angle changes from 0°to 90°.

In addition, Table (**1**), presents the thermophysical properties of water (H_2O) as the base fluid and magnetite (Fe_3O_4) as the nanofluid nanoparticle. Furthermore, it presented both the graphene (*GNP*) and titanium dioxide (TiO_2) as hybrid nanofluid nanoparticles. All the above, thermophysical properties were taken at the reference case of 298 *K*. Also, Table (**2**) represents the thermophysical properties specific to the spherical nanoparticle form of nanofluid and hybrid nanofluid. Restricted by the BCs [26]:

$$t = 0 : v_1 = 0, \quad v_2 = 0, \quad v_3 = 0, \quad T = T_0, \quad C = C_0, \tag{6}$$

$$t \geq 0 : v_1 = az_1, \quad v_2 = bz_2, \quad v_3 = 0, \quad T = T_0, \quad C = C_0, \quad as \quad z_3 = 0, \tag{7}$$

$$t \geq 0 : v_1 \to 0, \quad v_2 \to 0, \quad v_3 \to 0, \quad T \to T_\infty, \quad C \to C_\infty, \quad as \quad z_3 \to \infty. \tag{8}$$

Table 1. The titanium dioxide, graphene, magnetite, and water thermophysical properties [27-32].

Physical properties	Base fluid	Nanofluid nanoparticle	Hybrid nanofluid nanoparticles	
	Water (H_2O)	Magnetite (Fe_3O_4)	Graphene (GNP)	Titanium dioxide (TiO_2)
$\rho\,(kg\,m^{-3})$	997.1	5180	2250	4250
$\kappa\,(W\,m^{-1}\,K^{-1})$	0.613	9.7	2500	8.9538
$c_p\,(J\,kg^{-1}\,K^{-1})$	4179	670	2100	686.2
$\sigma\,(S\,m^{-1})$	0.05	25×10^3	1.0×10^7	2.38×10^6

(Table 1) cont.....

$\beta\,(K^{-1})$	2.57×10^{-4}	10.4×10^{-6}	2.1×10^{-6}	7.1×10^{-6}

Table 2. Thermophysical properties of nanoparticle-laden fluids [33-35].

Properties	Nanofluid	Hybrid nanofluid
Heat capacity	$(1-\phi_{s1})(\rho c_p)_{bf} + \phi_{s1}(\rho c_p)_{s1}$	$(1-\phi_{s2})(\rho c_p)_{nf} + \phi_{s2}(\rho c_p)_{s2}$
Thermal expansion	$(1-\phi_{s1})(\rho\beta)_{bf} + \phi_{s1}(\rho\beta)_{s1}$	$(1-\phi_{s2})(\rho\beta)_{nf} + \phi_{s2}(\rho\beta)_{s2}$
Density	$(1-\phi_{s1})\rho_{bf} + \phi_{s1}\rho_{s_1}$	$(1-\phi_{s2})\rho_{nf} + \phi_{s2}\rho_{s2}$
Mass diffusivity	$\frac{D_{bf}}{(1-\phi_{s1})^{2.5}}$	$\frac{D_{nf}}{(1-\phi_{s2})^{2.5}}$
Dynamic viscosity	$\frac{\mu_{bf}}{(1-\phi_{s1})^{2.5}}$	$\frac{\mu_{nf}}{(1-\phi_{s2})^{2.5}}$
Electrical conductivity	$\left(\frac{\sigma_{s_1} - 2\phi_{s1}(\sigma_{bf} - \sigma_{s1}) + 2\,\sigma_{bf}}{\sigma_{s_1} + \phi_{s1}(\sigma_{bf} - \sigma_{s1}) + 2\,\sigma_{bf}}\right)\sigma_{bf}$	$\left(\frac{\sigma_{s2} - 2\,\phi_{s2}(\sigma_{nf} - \sigma_{s2}) + 2\sigma_{nf}}{\sigma_{s2} + \phi_{s2}(\sigma_{nf} - \sigma_{s2}) + 2\sigma_{nf}}\right)\sigma_{nf}$
Thermal conductivity	$\left(\frac{\kappa_{s1} - 2\,\phi_{s1}(\kappa_{bf} - \kappa_{s1}) + 2\,\kappa_{bf}}{\kappa_{s1} + \phi_{s1}(\kappa_{bf} - \kappa_{s1}) + 2\,\kappa_{bf}}\right)\kappa_{bf}$	$\left(\frac{\kappa_{s2} - 2\,\phi_{s2}(\kappa_{nf} - \kappa_{s2}) + 2\kappa_{nf}}{\kappa_{s2} + \phi_{s2}(\kappa_{nf} - \kappa_{s2}) + 2\,\kappa_{nf}}\right)\kappa_{nf}$

The flow Eqs. (1)-(5) transformed into a non-dimensional form using a set of variables

$$v_1 = a\, z_1 \frac{\partial f(\zeta,\eta)}{\partial \eta}, \quad v_2 = a\, z_2 \frac{\partial h(\zeta,\eta)}{\partial \eta}, \quad v_3 = -\sqrt{a\, v_{bf}\, \zeta}\,[f(\zeta,\eta) + h(\zeta,\eta)], \qquad (9)$$

$$\eta = \sqrt{\frac{a}{\zeta v_{bf}}}\, z_3, \quad \zeta = 1 - e^{-\tau}, \quad \psi = \frac{T - T_\infty}{T_0 - T_\infty}, \quad \chi = \frac{C - C_\infty}{C_0 - C_\infty}, \quad \tau = a\,t. \qquad (10)$$

Implementing the similarity transformation Eqs. (9)-(10) into Eqs. (1)-(5) along with the BCs Eqs. (6)-(8), a set of N-PDEs are obtained.

$$(\zeta-\zeta^2)\frac{\partial f'}{\partial \zeta}=A_1 f'''-\frac{\eta}{2}(\zeta-1)f''+\zeta(f f''+h f''+A_2 M^2(h'\sin\theta\cos\theta-f'sin^2\theta)-f'^2+2\Omega h'+A_3 Ri\,\psi), \quad \textbf{(11)}$$

$$(\zeta-\zeta^2)\frac{\partial h'}{\partial \zeta}=A_1 h'''-\frac{\eta}{2}(\zeta-1)h''+\zeta(f h''+h h''+A_2 M^2(f'sin\,\theta\, cos\,\theta-h'sin^2\theta)-h'^2+2\Omega f'), \quad \textbf{(12)}$$

$$(\zeta-\zeta^2)\frac{\partial \psi}{\partial \zeta}=A_4\frac{1}{Pr}\psi''-\frac{\eta}{2}(\zeta-1)\psi'+\zeta(f\psi'+h\psi'+\Gamma\psi), \quad \textbf{(13)}$$

$$(\zeta-\zeta^2)\frac{\partial \chi}{\partial \zeta}=A_5\frac{1}{Sc}\chi''-\frac{\eta}{2}(\zeta-1)\chi'+\zeta(f\chi'+h\chi') \quad \textbf{(14)}$$

where $A_1, A_2, A_3, A_4,$ and A_5 are constants:

$$A_1=\frac{\mu_{hnf}/\mu_{bf}}{\rho_{hnf}/\rho_{bf}},\quad A_2=\frac{\sigma_{hnf}/\sigma_{bf}}{\rho_{hnf}/\rho_{bf}},\quad A_3=\frac{\beta_{hnf}}{\beta_{bf}},\quad A_4=\frac{\kappa_{hnf}/\kappa_{bf}}{(\rho c_p)_{hnf}/(\rho c_p)_{bf}},\quad A_5=\frac{D_{hnf}}{D_{bf}}.$$

The symbol (') indicates that the differentiation is done with respect to the variable η. The BCs are:

$$f(\zeta,0)=0,\quad f'(\zeta,0)=1,\quad h(\zeta,0)=0,\quad h'(\zeta,0)=\delta,\quad \psi(\zeta,0)=1,\quad \chi(\zeta,0)=1, \zeta\in[0,1),$$

$$f(\zeta,\eta)\to 0,\quad h(\zeta,\eta)\to 0,\quad \psi(\zeta,\eta)\to 0,\quad \chi(\zeta,\eta)\to 0,\quad \zeta\in[0,1),\quad as\quad \eta\to\infty. \quad \textbf{(15)}$$

The symbol δ represents stretching and has a value greater than zero. Other parameters in the context are defined as follows: Ω is the rotation parameter, M is the magnetic parameter, and Γ is a parameter that can either represent heat absorption (when it's less than zero) or heat generation (when it's greater than zero).

$$\delta=\frac{b}{a},\quad \Omega=\frac{\Omega_0}{a},\quad M^2=\frac{B_0^2\,\sigma_{bf}}{a\,\rho_{bf}},\quad \Gamma=\frac{Q_0}{a\,(\rho\, c_p)_{bf}}.$$

Also, Ri is the local Richardson number, Gr is the local Grashof number, Re is the local Reynolds number, Pr is the Prandtl number, and Sc is the Schmidt number, that are defined as:

$$Ri=\frac{Gr}{Re^2},\quad Gr=\frac{g\,\Delta T\,\beta_{bf}\,z_1^3}{\nu_{bf}^2},\quad Re=\frac{a\,z_1^2}{\nu_{bf}},\quad Pr=\frac{\nu_{bf}\,(\rho\, c_p)_{bf}}{\kappa_{bf}},\quad Sc=\frac{\nu_{bf}}{D_{bf}}.$$

The physical quantities being investigated pertaining to the skin friction coefficients along z_1 and z_2 axis are C_{fz_1}, C_{fz_2}. Also, Nu_{z_1}, and Sh_{z_1} are the local Nusselt number, and Sherwood number. We obtain the dimensionless version of these quantities.

$$Re^{\frac{1}{2}}\zeta^{\frac{1}{2}}C_{fz_1} = \frac{\mu_{hnf}}{\mu_{bf}} f''(\zeta,0), \quad Re^{\frac{1}{2}}\zeta^{\frac{1}{2}}C_{fz_2} = \frac{\mu_{hnf}}{\mu_{bf}} h''(\zeta,0),$$

$$Re^{-\frac{1}{2}}\zeta^{\frac{1}{2}}Nu_{z_1} = \frac{-\kappa_{hnf}}{\kappa_{bf}} \psi'(\zeta,0), \quad Re^{-\frac{1}{2}}\zeta^{\frac{1}{2}}Sh_{z_1} = \frac{-D_{hnf}}{D_{bf}} \chi'(\zeta,0).$$

NUMERICAL SOLUTION AND VALIDATION

The system of N-PDEs Eqs. (11)-(14), was solved numerically with BCs (15) by using the OMD-BSSIM. We can express the iterative scheme for OMD-SSIM method as follows:

SIMPLE ITERATION METHOD

The linearization and decoupling technique was employed, based on the simple iteration method, to linearize all the flow Eqs. (11)-(14) about one dependent variable at a time. The equations were linearized in the following order f, h, ψ, and χ, which resulted in iterative schemederived as,

$$\alpha_{10,n} f'''_{n+1} + \alpha_{11,n} f''_{n+1} + \alpha_{12,n} f'_{n+1} + \alpha_{13,n} f_{n+1} + \alpha_{14,n} \frac{\partial f'_{n+1}}{\partial \zeta} = R_{1,n}, \quad \textbf{(16)}$$

$$\alpha_{20,n} h'''_{n+1} + \alpha_{21,n} h''_{n+1} + \alpha_{22,n} h'_{n+1} + \alpha_{23,n} h_{n+1} + \alpha_{24,n} \frac{\partial h'_{n+1}}{\partial \zeta} = R_{2,n}, \quad \textbf{(17)}$$

$$\alpha_{30,n} \psi''_{n+1} + \alpha_{31,n} \psi'_{n+1} + \alpha_{32,n} \psi_{n+1} + \alpha_{33,n} \frac{\partial \psi_{n+1}}{\partial \zeta} = R_{3,n}, \quad \textbf{(18)}$$

$$\alpha_{40,n} \chi''_{n+1} + \alpha_{41,n} \chi'_{n+1} + \alpha_{42,n} \chi_{n+1} + \alpha_{43,n} \frac{\partial \chi_{n+1}}{\partial \zeta} = R_{4,n}, \quad \textbf{(19)}$$

where the coefficients and right-hand sides for Eqs. (16)-(19) are known functions at a particular iteration n. In the SIM implementation method, nonlinear terms are expressed as a combination of known functions at iteration n and unknown functions at iteration $n+1$. The unknown functions are determined by selecting the nonlinear term with the highest derivative in each Eq. [22]. This approach enables the calculation of the SIM coefficients and right-hand sides for each equation as:

$$\alpha_{10,n} = A_1, \quad \alpha_{20,n} = A_1, \quad \alpha_{30,n} = A_4 \frac{1}{Pr}, \quad \alpha_{40,n} = A_5 \frac{1}{Sc}, \quad \alpha_{11,n} = \frac{\eta}{2}(1-\zeta) + \zeta(f_n + h_n),$$

$$\alpha_{21,n} = \frac{\eta}{2}(1-\zeta) + \zeta(f_n + h_n), \quad \alpha_{31,n} = \frac{\eta}{2}(1-\zeta) + \zeta(f_n + h_n), \quad \alpha_{41,n} = \frac{\eta}{2}(1-\zeta) + \zeta(f_n + h_n),$$

$$\alpha_{12,n} = -\zeta(A_1\lambda + A_2M^2 sin^2\theta + f_n'), \quad \alpha_{22,n} = -\zeta(A_1\lambda + A_2M^2 sin^2\theta + h_n'), \quad \alpha_{32,n} = \Gamma\zeta, \alpha_{42,n} = 0,$$

$$\alpha_{13,n} = 0, \quad \alpha_{23,r} = 0, \quad \alpha_{33,n} = \zeta^2 - \zeta, \quad \alpha_{43,n} = \zeta^2 - \zeta, \quad \alpha_{14,n} = \zeta^2 - \zeta, \alpha_{24,n} = \zeta^2 - \zeta, R_{3,n} = 0,$$

$$R_{1,n} = -(2\Omega + A_2M^2 \sin\theta\cos\theta)h_n' - A_3\, Ri\, \psi_n, \quad R_{2,n} = (2\Omega + A_2M^2 \sin\theta\cos\theta)f_{n+1}', R_{4,n} = 0.$$

CHEBYSHEV DIFFERENTIATION

The overlapping multi-domain spectral collocation method utilizes an overlapping grid procedure to divide the space variable (η) domain into smaller subdomains of uniform length, before the implementation of the spectral method. In contrast, the time variable (ζ) domain is divided into smaller sub-intervals that are not overlapping. This approach is designed to improve the efficiency and accuracy of the spectral collocation method. Fig. (**2**) shows the non-overlapping time ζ domain I, with $I \in [\zeta_0, \zeta_F]$. The ζ domain is partitioned into p non-overlapping sub-intervals of equal length defined as:

$$I_\varepsilon = [\zeta_{\varepsilon-1}, \zeta_\varepsilon], \quad \zeta_{\varepsilon-1} < \zeta_\varepsilon, \quad \varepsilon = 1,2,3,4, \dots, p.$$

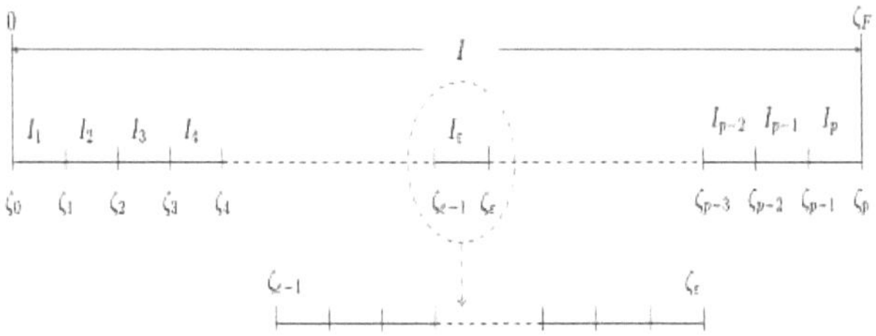

Fig. (2). Non-overlapping grid (ζ domain).

Fig. (**3**) displays the division of the overlapping space η domain Υ, with $\Upsilon \in [0, \eta_\infty]$. The η domain is partitioned into q overlapping sub-intervals, which are specified by the following definition:

$$\Upsilon_{\varrho} = \left[\eta_0^{\varrho}, \eta_{N_\eta}^{\varrho}\right], \quad \varrho = 1,2,3,\ldots,q.$$

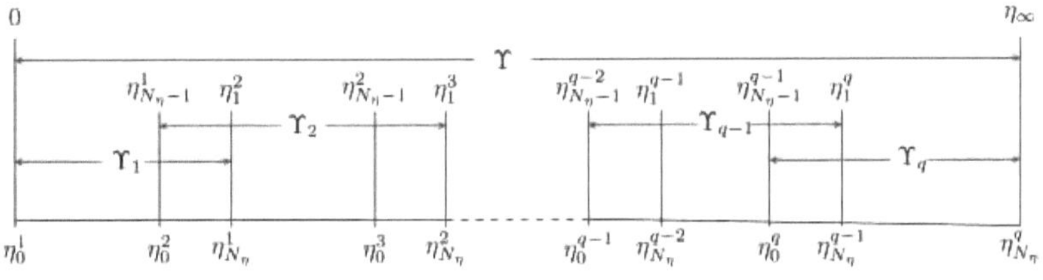

Fig. (3). *Overlapping grid (η domain).*

Before using the spectral collocation method, it is a common practice to transform the time interval I_ε and space interval Υ_ϱ into the interval of [-1, 1] through linear transformation mapping [20]. The solution procedure assumes that the solution can be approximated by a bivariate Lagrange interpolating polynomial form [36],

$$\overset{(\varrho)}{f}(\zeta,\eta) \approx \sum_{i=0}^{N_\eta}\sum_{j=0}^{N_\zeta} \overset{(\varrho)}{F}(\hat{\zeta}_j,\hat{\eta}_i)L_i(\eta)L_j(\zeta),\ \overset{(\varrho)}{h}(\zeta,\eta) \approx \sum_{i=0}^{N_\eta}\sum_{j=0}^{N_\zeta} \overset{(\varrho)}{H}(\hat{\zeta}_j,\hat{\eta}_i)L_i(\eta)L_j(\zeta), \quad (20)$$

$$\overset{(\varrho)}{\psi}(\zeta,\eta) \approx \sum_{i=0}^{N_\eta}\sum_{j=0}^{N_\zeta} \overset{(\varrho)}{\psi}(\hat{\zeta}_j,\hat{\eta}_i)L_i(\eta)L_j(\zeta),\ \overset{(\varrho)}{\chi}(\zeta,\eta) \approx \sum_{i=0}^{N_\eta}\sum_{j=0}^{N_\zeta} \overset{(\varrho)}{X}(\hat{\zeta}_j,\hat{\eta}_i)L_i(\eta)L_j(\zeta), \quad (21)$$

which interpolate as selected points $\hat{\eta}_i$ and $\hat{\zeta}_j$. The functions $L_i(\eta)$ and $L_j(\zeta)$ are the characteristic Lagrange cardinal polynomial [37]. From Eqs. (20)-(21), the approximated first spatial derivatives are computed as:

$$\left.\frac{\partial \overset{(\varepsilon)}{f}}{\partial \eta}\right|_{(\hat{\zeta}_i,\hat{\eta}_j)} = \boldsymbol{D}\overset{(\varepsilon)}{\boldsymbol{F}}_j,\ \left.\frac{\partial \overset{(\varepsilon)}{h}}{\partial \eta}\right|_{(\hat{\zeta}_i,\hat{\eta}_j)} = \boldsymbol{D}\overset{(\varepsilon)}{\boldsymbol{H}}_j,\ \left.\frac{\partial \overset{(\varepsilon)}{\psi}}{\partial \eta}\right|_{(\hat{\zeta}_i,\hat{\eta}_j)} = \boldsymbol{D}\overset{(\varepsilon)}{\boldsymbol{\psi}}_j,\ \left.\frac{\partial \overset{(\varepsilon)}{\chi}}{\partial \eta}\right|_{(\hat{\zeta}_i,\hat{\eta}_j)} = \boldsymbol{D}\overset{(\varepsilon)}{\boldsymbol{X}}_j, \quad (22)$$

where $\boldsymbol{D}$ represents the Chebyshev spatial differentiation matrix in the ϱth sub-domain as defined in [19]. The matrix $\boldsymbol{D}$ is size $(S+1)\times(S+1)$, where $S = N_\eta + (q-1)(N_\eta - 1)$ is the sum of all grid points in the whole space domain. The vector is defined as:

$$\overset{(\varepsilon)}{\boldsymbol{F}_j} = \left[\overset{(\varepsilon)}{f}\left(\hat{\zeta}_j,\overset{(\varrho)}{\hat{\eta}_0}\right), \overset{(\varepsilon)}{f}\left(\hat{\zeta}_j,\overset{(\varrho)}{\hat{\eta}_1}\right), \overset{(\varepsilon)}{f}\left(\hat{\zeta}_j,\overset{(\varrho)}{\hat{\eta}_2}\right), \ldots, \overset{(\varepsilon)}{f}\left(\hat{\zeta}_j,\overset{(\varrho)}{\hat{\eta}_{N_\eta}}\right)\right]^{t'},$$

With t' represents the matrix transpose. Similarly, we define for $\overset{(\varepsilon)}{\boldsymbol{H}_j}$, $\overset{(\varepsilon)}{\boldsymbol{\psi}_j}$, and $\overset{(\varepsilon)}{\boldsymbol{X}_j}$. The higher order (s) spatial derivatives were obtained as powers of $\boldsymbol{D}$, which are:

$$\left.\frac{\partial^s \overset{(\varepsilon)}{f}}{\partial \eta^s}\right|_{(\zeta_\iota,\eta_j)} = D^s \overset{(\varepsilon)}{F_j}, \quad \left.\frac{\partial^s \overset{(\varepsilon)}{h}}{\partial \eta^s}\right|_{(\zeta_\iota,\eta_j)} = D^s \overset{(\varepsilon)}{H_j}, \quad \left.\frac{\partial^s \overset{(\varepsilon)}{\psi}}{\partial \eta^s}\right|_{(\zeta_\iota,\eta_j)} = D^s \overset{(\varepsilon)}{\psi_j}, \quad \left.\frac{\partial^s \overset{(\varepsilon)}{\chi}}{\partial \eta^s}\right|_{(\zeta_\iota,\eta_j)} = D^s \overset{(\varepsilon)}{X_j}, \quad (23)$$

The approximated time derivative is estimated as,

$$\left.\frac{\partial \overset{(\varepsilon)}{f'}}{\partial \zeta}\right|_{(\zeta_\iota,\eta_j)} = \sum_{\iota=0}^{N_\zeta} d_{j,\iota}\, D\overset{(\varepsilon)}{F_j}, \quad \left.\frac{\partial \overset{(\varepsilon)}{h'}}{\partial \zeta}\right|_{(\zeta_\iota,\eta_j)} = \sum_{\iota=0}^{N_\zeta} d_{j,\iota}\, D\overset{(\varepsilon)}{H_j}, \quad (24)$$

$$\left.\frac{\partial \overset{(\varepsilon)}{\psi}}{\partial \zeta}\right|_{(\zeta_\iota,\eta_j)} = \sum_{\iota=0}^{N_\zeta} d_{j,\iota}\, \overset{(\varepsilon)}{\psi_j}, \quad \left.\frac{\partial \overset{(\varepsilon)}{\chi}}{\partial \zeta}\right|_{(\zeta_\iota,\eta_j)} = \sum_{\iota=0}^{N_\zeta} d_{j,\iota}\, \overset{(\varepsilon)}{X_j}, \quad (25)$$

where $d_{j,\iota}$ being the standard first-order Chebyshev differentiation matrix of size $(N_\zeta + 1) \times (N_\zeta + 1)$. Substituting Eqs. (20)-(25) into Eqs. (16)-(19), we obtain the matrix equations:

$$\overset{(i)(\varepsilon)}{\widehat{\mathbf{A}}}\, \overset{(\varepsilon)}{\widehat{\mathbf{F}}} = \widehat{\mathbf{K}}_1, \quad \overset{(i)(\varepsilon)}{\widehat{\mathbf{B}}}\, \overset{(\varepsilon)}{\widehat{\mathbf{H}}} = \widehat{\mathbf{K}}_2, \quad \overset{(i)(\varepsilon)}{\widehat{\mathbf{C}}}\, \overset{(\varepsilon)}{\widehat{\boldsymbol{\psi}}} = \widehat{\mathbf{K}}_3, \quad \overset{(i)(\varepsilon)}{\widehat{\mathbf{E}}}\, \overset{(\varepsilon)}{\widehat{\mathbf{X}}} = \widehat{\mathbf{K}}_4. \quad (26)$$

Equation (26) **is** expressed as the $N_\zeta(S+1) \times N_\zeta(S+1)$ matrix system, and the vectors $\widehat{\mathbf{F}}$, $\widehat{\mathbf{H}}$, $\widehat{\boldsymbol{\psi}}$, and $\widehat{\mathbf{X}}$ are size $S+1$, defined as:

$$\overset{(i)}{\widehat{\mathbf{A}}} = \overset{(\varepsilon)}{\alpha_{10,n}}\, D^3 + \overset{(\varepsilon)}{\alpha_{11,n}}\, D^2 + \overset{(\varepsilon)}{\alpha_{12,n}}\, D + \overset{(\varepsilon)}{\alpha_{13,n}}\, \mathbf{I}, \quad \overset{(\varepsilon)}{\widehat{\mathbf{K}}_1} = \overset{(\varepsilon)}{\widehat{R}}_{1,i,n} - \overset{(\varepsilon)}{\alpha_{14,n}}\, d_{i,N_\zeta}\, D\overset{(\varepsilon)}{\widehat{F}}_{N_\zeta,n+1},$$

$$\overset{(i)}{\widehat{\mathbf{B}}} = \overset{(\varepsilon)}{\alpha_{20,n}}\, D^3 + \overset{(\varepsilon)}{\alpha_{21,n}}\, D^2 + \overset{(\varepsilon)}{\alpha_{22,n}}\, D + \overset{(\varepsilon)}{\alpha_{23,n}}\, \mathbf{I}, \quad \overset{(\varepsilon)}{\widehat{\mathbf{K}}_2} = \overset{(\varepsilon)}{\widehat{R}}_{2,i,n} - \overset{(\varepsilon)}{\alpha_{14,n}}\, d_{i,N_\zeta}\, D\overset{(\varepsilon)}{\widehat{H}}_{N_\zeta,n+1},$$

$$\overset{(i)}{\widehat{\mathbf{C}}} = \overset{(\varepsilon)}{\alpha_{30,n}}\, D^2 + \overset{(\varepsilon)}{\alpha_{31,n}}\, D + \overset{(\varepsilon)}{\alpha_{32,n}}\, \mathbf{I}, \quad \overset{(\varepsilon)}{\widehat{\mathbf{K}}_3} = \overset{(\varepsilon)}{\widehat{R}}_{3,i,n} - \overset{(\varepsilon)}{\alpha_{33,n}}\, d_{i,N_\zeta}\, \overset{(\varepsilon)}{\widehat{F}}_{N_\zeta,n+1},$$

$$\overset{(i)}{\widehat{\mathbf{E}}} = \overset{(\varepsilon)}{\alpha_{40,n}}\, D^2 + \overset{(\varepsilon)}{\alpha_{41,n}}\, D + \overset{(\varepsilon)}{\alpha_{42,n}}\, \mathbf{I}, \quad \overset{(\varepsilon)}{\widehat{\mathbf{K}}_4} = \overset{(\varepsilon)}{\widehat{R}}_{4,i,n} - \overset{(\varepsilon)}{\alpha_{43,n}}\, d_{i,N_\zeta}\, \overset{(\varepsilon)}{\widehat{F}}_{N_\zeta,n+1}.$$

Where $\mathbf{I}$ is the identity matrix. Subject to the BCs:

$$\overset{(\varepsilon)}{F}_{n+1}(\zeta_i,\eta_S)=0,\ \ \sum_{k=0}^{S}\mathbf{D}_{S,k}\overset{(\varepsilon)}{F}_{n+1}(\zeta_i,\eta_k)=1,\ \ \sum_{k=0}^{S}\mathbf{D}_{0,k}\overset{(\varepsilon)}{F}_{n+1}(\zeta_i,\eta_k)=0,\quad \textbf{(27)}$$

$$\overset{(\varepsilon)}{H}_{n+1}(\zeta_i,\eta_S)=0,\ \ \sum_{k=0}^{S}\mathbf{D}_{S,k}\overset{(\varepsilon)}{H}_{n+1}(\zeta_i,\eta_k)=\delta,\ \ \sum_{k=0}^{S}\mathbf{D}_{0,k}\overset{(\varepsilon)}{H}_{n+1}(\zeta_i,\eta_k)=0,\quad \textbf{(28)}$$

$$\overset{(\varepsilon)}{\psi}_{n+1}(\zeta_i,\eta_S)=1,\ \ \overset{(\varepsilon)}{\psi}_{n+1}(\zeta_i,\eta_0)=0,\ \overset{(\varepsilon)}{X}_{n+1}(\zeta_i,\eta_S)=1,\ \ \overset{(\varepsilon)}{X}_{n+1}(\zeta_i,\eta_0)=0.\quad \textbf{(29)}$$

The functions taken as guesses for initiating the iteration procedure which satisfies the BCs Eqs. (27)-(29).

$$\overset{(\varepsilon)}{F}_{j,0}=1-e^{-\eta},\overset{(\varepsilon)}{H}_{j,0}=\delta(1-e^{-\eta}),\overset{(\varepsilon)}{\psi}_{j,0}=e^{-\eta},\overset{(\varepsilon)}{X}_{j,0}=e^{-\eta}.$$

After imposing the BCs Eqs. (27)-(29) into the matrix system Eq. (26), the solutions are derived by solving the resulting matrix systems iteratively.

$$\overset{(\varepsilon)}{\hat{\mathbb{F}}}=\mathbf{inv}\left(\overset{(i)}{\hat{\mathbb{A}}}\right)\overset{(\varepsilon)}{\hat{\mathbb{R}}}_1,\ \ \overset{(\varepsilon)}{\hat{\mathbb{H}}}=\mathbf{inv}\left(\overset{(i)}{\hat{\mathbb{B}}}\right)\overset{(\varepsilon)}{\hat{\mathbb{R}}}_2,\ \ \overset{(\varepsilon)}{\hat{\psi}}=\mathbf{inv}\left(\overset{(i)}{\hat{\mathbb{C}}}\right)\overset{(\varepsilon)}{\hat{\mathbb{R}}}_3,\ \ \overset{(\varepsilon)}{\hat{\mathbb{X}}}=\mathbf{inv}\left(\overset{(i)}{\hat{\mathbb{E}}}\right)\overset{(\varepsilon)}{\hat{\mathbb{R}}}_4.\quad \textbf{(30)}$$

Here **inv** represents the inverse of the matrix.

RESULT AND DISCUSSION

The system of N-PDEs is represented by Eqs. (11)-(14) along with the BCs Eq. (15), solved numerically using OMD-BSSIM. At the $n\,th$ iteration level and for time and space collocation nodes, the approximated solutions are denoted by $\boldsymbol{F}_{\mathrm{j,n}}$, $\boldsymbol{H}_{\mathrm{j,n}}$, $\boldsymbol{\psi}_{\mathrm{j,n}}$, and $\boldsymbol{X}_{\mathrm{j,n}}$. The numerical computations were executed using the provided values:

$$\phi_{s1}=0.2,\ \phi_{s2}=0.02,\ Ri=4,\ Pr=7,\ Sc=2,\ \delta=0.5,\ \Omega=2,$$

$$M=3, \Gamma=2,\ \theta=\pi/3, n=25,\ p=20,\ N_{\zeta}=5,\ \eta_{\infty}=10, \zeta_F=0.94.$$

Unless otherwise indicated, these values have remained unchanged throughout the entirety of the work. Typically, the formula for calculating the total number of grid points (S) in overlapping nodes with dimensions of $[N_{\eta}, q]$ was used. To determine the precision of the iterative OMD-BSSIM, we evaluate it by examining both the solution error and residual error. In this study, we established the solution errors to be:

$$E_f = \left\|\boldsymbol{F}_{\mathrm{j,n+1}} - \boldsymbol{F}_{\mathrm{j,n}}\right\|_{\infty}, E_h = \left\|\boldsymbol{H}_{\mathrm{j,n+1}} - \boldsymbol{H}_{\mathrm{j,n}}\right\|_{\infty},$$

$$E_{\psi} = \left\|\boldsymbol{\psi}_{\mathrm{j,n+1}} - \boldsymbol{\psi}_{\mathrm{j,n}}\right\|_{\infty}, E_{\chi} = \left\|\boldsymbol{X}_{\mathrm{j,n+1}} - \boldsymbol{X}_{\mathrm{j,n}}\right\|_{\infty}.$$

Fig. (**4**) displays how the error norms in the solution change as the number of iterations (n) increases, assuming an equal $S = 91$. The idea of using overlapping grids in spatial computations (Fig. **4a**) has been observed to improve efficiency and achieve better accuracy compared to non-overlapping grids (Fig. **4b**). It is also noted that convergence is fully achieved after approximately 10 iterations.

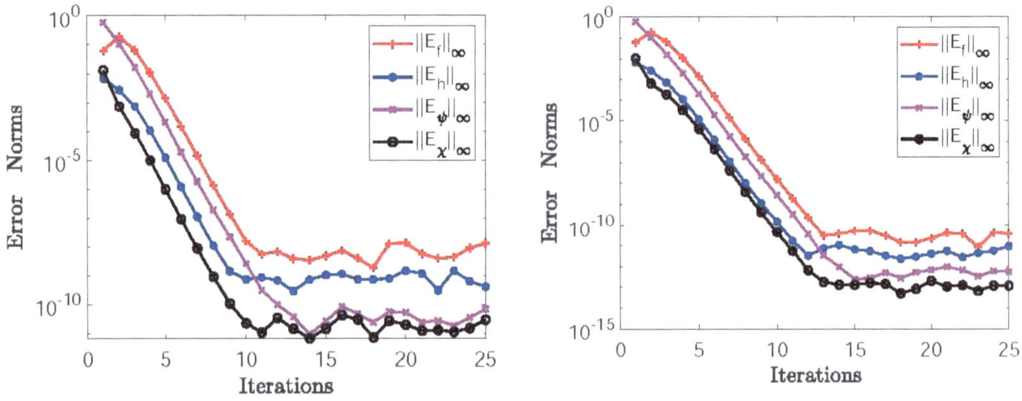

(a) Non-overlapping the space domain (N_{η} = 91, q = 1) (b) Overlapping the space domain (N_{η} = 16, q = 6)

Fig. (4). The solutions error graphs of the non-overlapping and overlapping.

The residual errors, on the other hand, are defined as the difference between the computed value of the N-PDEs equations (11)-(14), and the approximate solution at each point in the discretized domain. Fig. (**5**) presents the residual errors against space and time. It is noted that the calculated residual errors for the velocities, temperature, and concentration profiles (Figs. **5a** to **5d**) were $O(10^{-9})$, $O(10^{-10})$, $O(10^{-12})$, and $O(10^{-14})$, respectively.

(a) Residual error graph of $f'(\zeta, \eta)$.

(b) Residual error graph of $h'(\zeta, \eta)$.

(c) Residual error graph of $\vartheta(\zeta, \eta)$.

(d) Residual error graph of $\phi(\zeta, \eta)$.

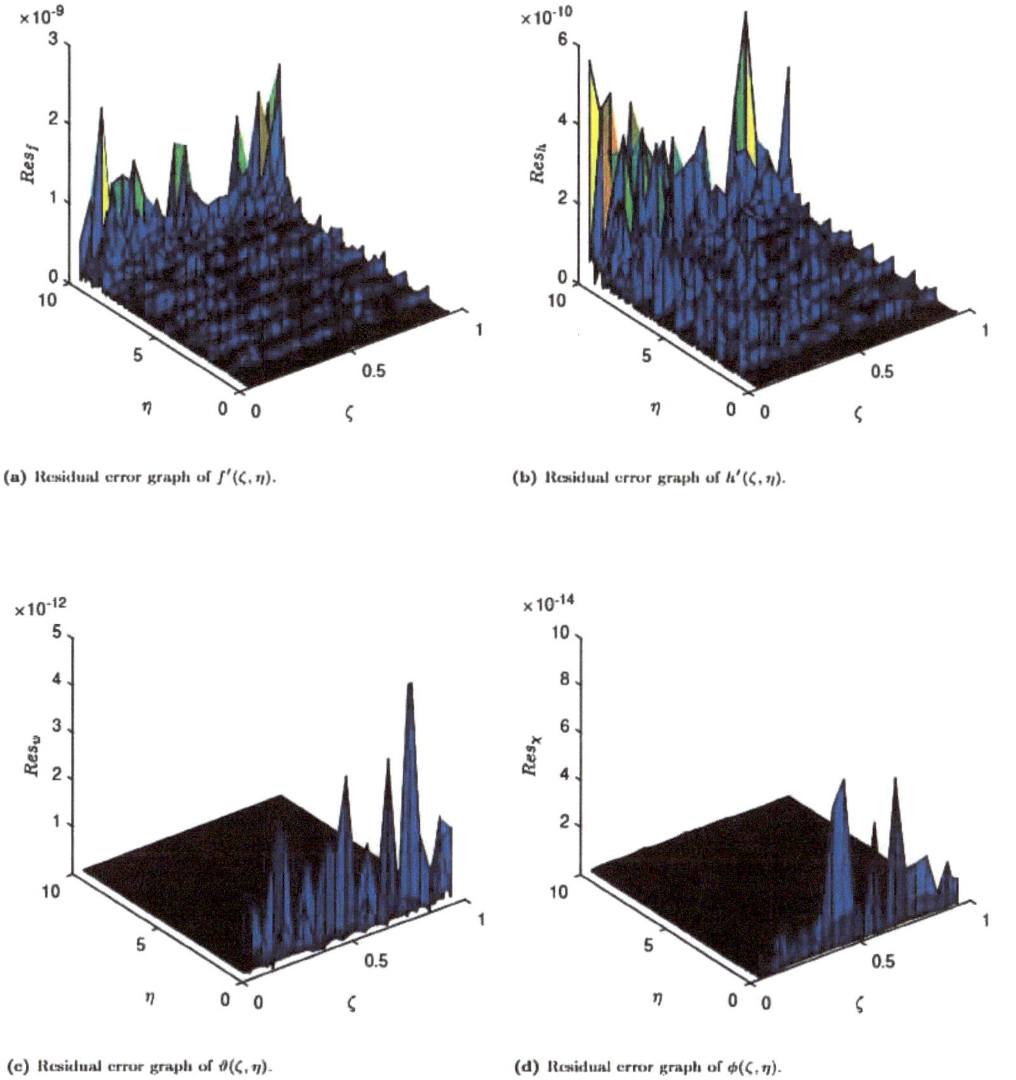

Fig. (5). The residual error graphs of the OMD-BSSIM solution when $q = 6$, and $N_\eta = 16$.

To validate the utilization of OMD-BSSIM, we restricted the possible values of the space node to $q = 3$ and $N_\eta = 14$, resulting in a total of $S = 40$ grid points for overlapping domains. For the time variable, we took different values of ζ_F while keeping $N_\zeta = 10$. In the absence of rotation, heat generation, and absorption parameters, Richardson number, and nanoparticle volume fraction of magnetite and graphene or titanium dioxide $(\Omega = \Gamma = Ri = \phi_{s1} = \phi_{s2} = 0)$, we kept $M =$

$1, Pr = 0.7, \delta = 0.5$, and $\theta = \pi/2$. Under these constraints, we solved the same model as in Dlamini *et al.* [38] using the compact finite difference relaxation method (CFDRM) and the Keller box method (KBM). The findings from their research paper have been combined with the latest results from the OMD-BSSIM and are presented in Table (**3**). The results indicate that using the overlapping grid technique significantly reduces computational time.

Table 3. Comparing the performance of OMD-BSSIM with CFDRM and KBM Methods in Computing $f''(\zeta, 0)$, and $h''(\zeta, 0)$.

ζ_F	$-f''(\zeta,0)$			$-h''(\zeta,0)$		
	OMD-BSSIM	CFDRM	KBM	OMD-BSSIM	CFDRM	KBM
0.1	0.67447	0.67444	0.6444	0.32760	0.32758	0.32758
0.3	0.88042	0.88040	0.88040	0.41457	0.41456	0.41456
0.5	1.06893	1.06893	1.06893	0.49637	0.49636	0.52099
0.7	1.24206	1.24206	1.24206	0.57322	0.57324	0.53404
0.9	1.40159	1.40159	1.40159	0.64538	0.64538	0.53712
CPU time	4.4108	6.047	238.769			

Fig. (**6**) illustrates the influence of two different nanoparticles (GNP and TiO_2) and various physical parameters, such as the Richardson number, volume fraction coefficients, angle, rotation, and stretching ratio parameters, on the temperature distribution $\psi(\zeta, \eta)$. As displayed in Figs. (**6a** and **6b**), adding nanoparticles to a fluid can enhance its thermal conductivity, resulting in a more uniform temperature profile in the boundary layer due to the high surface area-to-volume ratio of nanoparticles, which enhances the heat transfer rate. When a magnetic field is applied to a conducting fluid, it induces an electric current that interacts with the field, resulting in a Lorentz force that affects fluid flow and temperature distribution. As shown in Fig. (**6c**), applying the magnetic field at an angle to the fluid flow induces sthe econdary flows that enhance fluid mixing and reduce the thermal boundary layer, resulting in more efficient heat transfer and more uniform temperature distribution. The temperature profile of hybrid nanofluids is directly enhanced with increasing values of the rotation parameter (Ω), as depicted in Fig. (**6d**). Rotation has the potential to create additional mechanisms for transporting fluid, such as centrifugal forces, that affect the concentration gradients within the fluid, enhancing the process of diffusion. As illustrated in Fig. (**6e**), an increase in the stretching ratio parameter, particularly when stretching in the z_1 direction is greater than the z_2 direction (0.25 and 0.5) or when stretching in the z_2 direction is greater than the z_1 direction (2 and 4), leads to a decrease in the temperature profile due to a stronger velocity gradient that enhances fluid mixing and increases the thickness of the thermal boundary layer, resulting in a lower temperature gradient and a reduced temperature profile. Increasing the stretching ratio parameter in the z_2 direction can result in a more uniform temperature profile distribution but with a reduced heat transfer rate. Increasing the Richardson number between 1 to 4 indicates stronger buoyancy forces relative to viscous forces, impacting fluid flow patterns and temperature distribution in the boundary layer. As demonstrated in Fig. (**6f**), increasing the Richardson number can result in a more stable boundary layer with less mixing and less efficient heat transfer, due to the stronger buoyancy forces that create stratification and suppress fluid mixing.

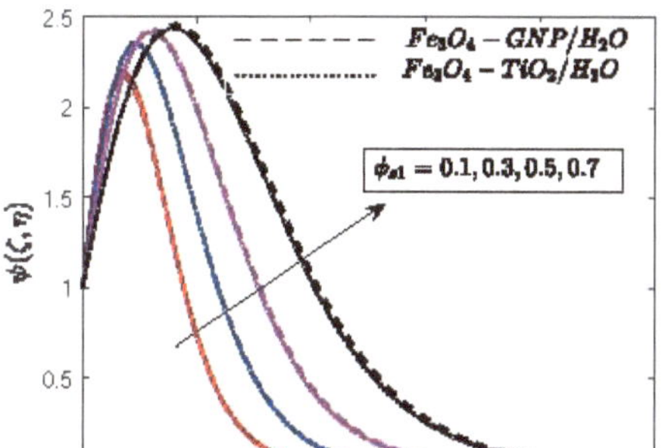

(a) Effect of ϕ_{s1}.

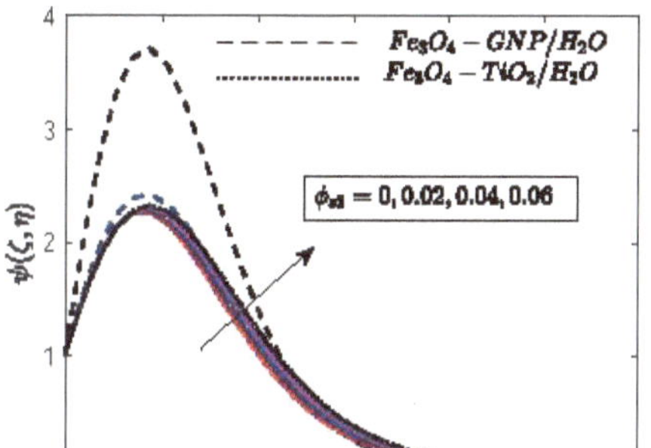

(b) Effect of ϕ_{s2}.

(Fig. 6) contd.....

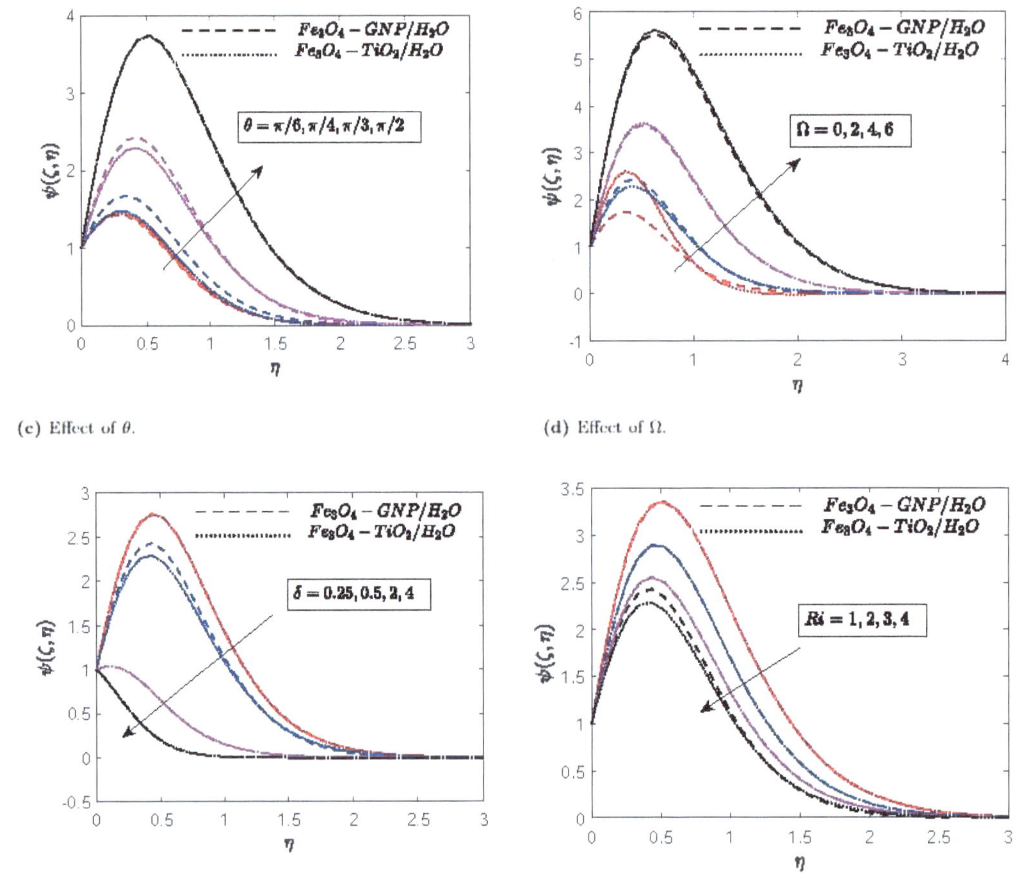

(c) Effect of θ.

(d) Effect of Ω.

(e) Effect of δ.

(f) Effect of Ri.

Fig. (6). Fluctuation of $\psi(\zeta,\eta)$ when $q = 6$,and $N_\eta = 16$.

In this study, we aimed to investigate the relationship between various parameters, including the stretching ratio parameter, projection angle, volume fraction coefficient of magnetite, and rotation parameter, and their impact on the skin friction coefficients, local Nusselt number, and Sherwood number in fluid flow. To accomplish this objective, we employed R visualization methods, particularly line graphs, to visualize the trends and patterns in the data, as shown in Fig. (7). The x-axis of the graph corresponded to the independent variables (δ, θ, ϕ_{S1}, and Ω), while the y-axis represented the dependent variables (C_{fz_1}, C_{fz_2}, Nu_{z_1}, and Sh_{z_1}).

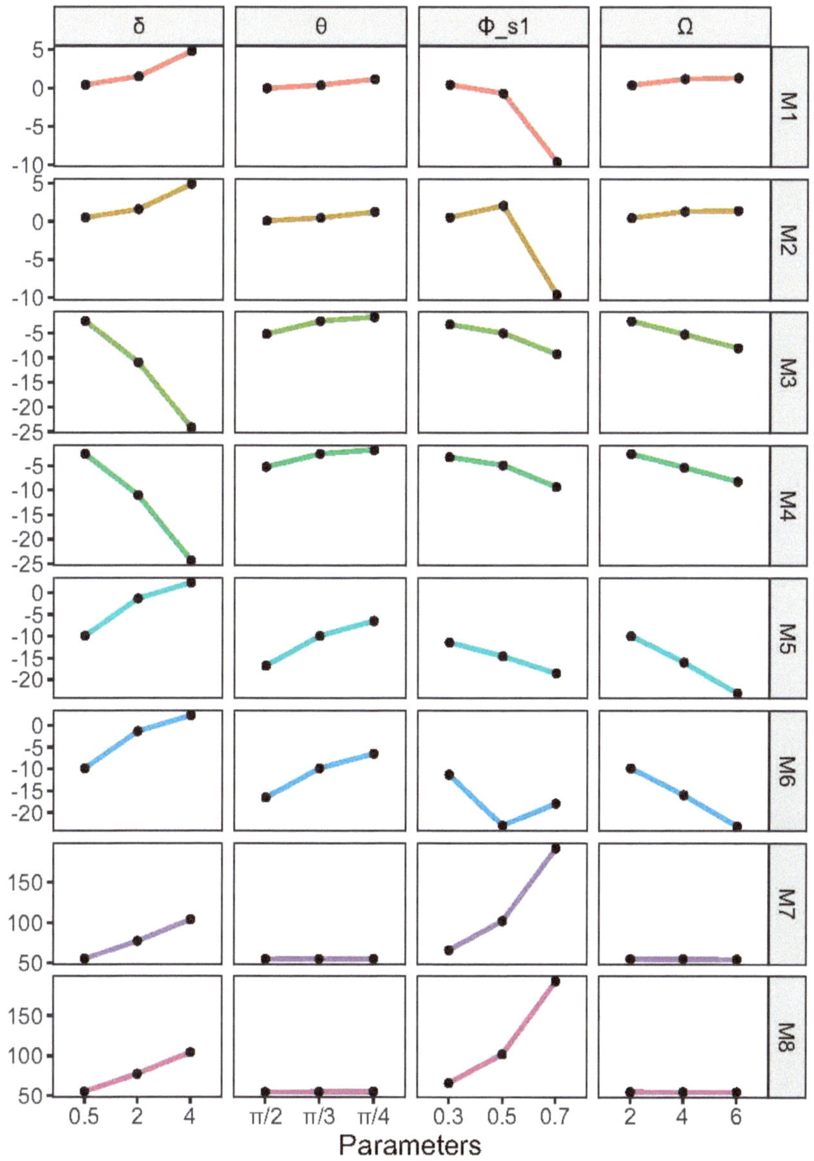

Fig. (7). Visualization of skin friction coefficient in z_1 direction, and in z_2 direction, local Nusselt number, and Sherwood number for the $GNP - Fe_3O_4/H_2O$, and $TiO_2 - Fe_3O_4/H_2O$ hybrid nanofluids.

M1: C_{fz_1} for the graphene hybrid nanofluid M2: C_{fz_1} for the graphene hybrid nanofluid
M3: C_{fz_2} for the graphene hybrid nanofluid M4: C_{fz_2} for the graphene hybrid nanofluid
M5: Nu_{z_1} for the graphene hybrid nanofluid M6: Nu_{z_1} for the graphene hybrid nanofluid
M7: Sh_{z_1} for the graphene hybrid nanofluid M8: Sh_{z_1} for the graphene hybrid nanofluid

Moreover, we conducted a regression analysis to determine the degree and direction of the relationship between the variables. Our findings demonstrated a statistically significant and positive association between the variable $\theta = \pi/4$ and multiple outcome measures, including M1, M2, M3, M4, M5, and M6. Additionally, notable positive relationships were observed between ϕ_{s1} and M6, as well as M8, and between Ω and M1 and M2. On the other hand, the variable δ had a negative impact on M3 and M4. These results emphasize the usefulness of R visualization techniques in revealing complex patterns and connections in intricate datasets related to fluid flow.

CONCLUSION

The main objective of this study is to investigate the heat transfer efficiency and flow characteristics of unsteady, three-dimensional, magnetohydrodynamic, incompressible, electrically conducting, rotating, and stratified hybrid nanofluid composed of Fe_3O_4/H_2O. We employed the OMD-BSSIM to solve the flow model and analyze the effects of various parameters on the temperature profile, skin friction coefficients, local Nusselt number, and Sherwood number of the hybrid nanofluid. By employing R visualization techniques and graphs, we were able to analyze the relationships between these parameters and the fluid flow characteristics. Our study contributes important insights into the behaviour of unsteady MHD hybrid nanofluid flow and its heat transfer efficiency. In conclusion, our study sheds light on the relationship between various parameters and their impact on fluid flow. The findings of this research can assist researchers and engineers in designing more efficient and effective fluid flow systems. Future research can explore the effect of other parameters and their interplay with the parameters examined in this study to gain a more comprehensive understanding of fluid flow.

LIST OF NOMENCLATURE, SUBSCRIPTS, ABBREVIATIONS AND GREEK SYMBOLS

Nomenclature			
z_1, z_2, z_3	Cartesian coordinate system	κ Ω	Thermal conductivity Rotation parameter
v_1, v_2, v_3	Velocities in direction z_1, z_2, z_3	ϕ	Nanoparticle volume fraction
t	Time variable	δ	Stretching ratio parameter
T	Fluid temperature	η	Coordinate for space transformation

cont.....

T_∞	Ambient temperature	ζ	Coordinate for time transformation
T_0	Wall temperature	ψ	Non-dimensional temperature
C	Fluid concentration	χ	Non-dimensional concentration
C_∞	Ambient concentration	**Subscripts**	
C_0	Wall concentration	bf	Base fluid
B_0	Magnetic field coefficient	nf	Nanofluid
Q_0	Heat generation/absorption coefficient	hnf	Hybrid nanofluid
Gr	Grashof number	$s1$	Magnetite nanoparticle
Re	Reynolds number	$s2$	Graphene or titanium dioxide nanoparticles
Ri	Richardson number	**Abbreviations**	
Pr	Prandtl number	MHD	Magnetohydrodynamic
c_p	Capacity of specific heat	BCs	Boundary conditions
f, h	Velocity components after transformation	SIM	Simple iteration method
C_{fz_1}, C_{fz_2}	Skin friction coefficients	BSRM	Bivariate spectral relaxation method
Nu_{z_1}	Nusselt number	OMD	Overlapping grid multi-domain
Sh_{z_1}	Sherwood number	BSSIM	Bivariate spectral simple iteration method
Greek Symbols		-	-
θ	Magnetic angle between z_1, and z_2 axis	-	- 978-981-5223-70-5

cont.....

Γ	Heat generation/absorption parameter	-	-
μ	Dynamic viscosity	-	-
ρ	Density	-	-
σ	Electrical conductivity	-	-
Ω_0	Angular velocity	-	-

REFERENCES

[1] B.C. Sakiadis, "Boundary layer behavior on continuous solid surfaces: I. Boundary-layer equations for two-dimensional and axisymmetric flow", *AIChE J.,* vol. 7, no. 1, pp. 26-28, 1961.
http://dx.doi.org/10.1002/aic.690070108

[2] B.C. Sakiadis, "Boundary layer behavior on continuous solid surfaces: II. The boundary layer on a continuous flat surface", *AIChE J.,* vol. 7, no. 2, pp. 221-225, 1961.
http://dx.doi.org/10.1002/aic.690070211

[3] L.J. Crane, "Flow past a stretching plate", *Zeitschrift für angewandte Mathematik und Physik ZAMP,,* vol. 21, pp. 645-647, 1970.
http://dx.doi.org/10.1007/BF01587695

[4] M.K. Laha, P.S. Gupta, and A.S. Gupta, "Heat transfer characteristics of the flow of an incompressible viscous fluid over a stretching sheet", *Wärme- Stoffübertrag.,* vol. 24, no. 3, pp. 151-153, 1989.
http://dx.doi.org/10.1007/BF01590013

[5] L.J. Grubka, and K.M. Bobba, "Heat transfer characteristics of a continuous stretching surface with variable temperature", *J. Heat Transfer,* vol. 107, no. 1, pp. 248-250, 1985.
http://dx.doi.org/10.1115/1.3247387

[6] S.U. Choi, and J.A. Eastman, "Enhancing thermal conductivity of fluids with nanoparticles," *Argonne National Lab.,* ANL: Argonne, IL, United States, 1995.

[7] Y. Li, J. Zhou, S. Tung, E. Schneider, and S. Xi, "A review on development of nanofluid preparation and characterization", *Powder Technol.,* vol. 196, no. 2, pp. 89-101, 2009.
http://dx.doi.org/10.1016/j.powtec.2009.07.025

[8] N. Sezer, M.A. Atieh, and M. Koç, "A comprehensive review on synthesis, stability, thermophysical properties, and characterization of nanofluids", *Powder Technol.,* vol. 344, pp. 404-431, 2019.
http://dx.doi.org/10.1016/j.powtec.2018.12.016

[9] G. Huminic, and A. Huminic, "Entropy generation of nanofluid and hybrid nanofluid flow in thermal systems: A review", *J. Mol. Liq.*, vol. 302, p. 112533, 2020.
http://dx.doi.org/10.1016/j.molliq.2020.112533

[10] M. Mehrali, E. Sadeghinezhad, A.R. Akhiani, S. Tahan Latibari, H.S.C. Metselaar, A.S. Kherbeet, and M. Mehrali, "Heat transfer and entropy generation analysis of hybrid graphene/Fe_3O_4 ferro-nanofluid flow under the influence of a magnetic field", *Powder Technol.*, vol. 308, pp. 149-157, 2017.
http://dx.doi.org/10.1016/j.powtec.2016.12.024

[11] J. Sarkar, "A critical review on convective heat transfer correlations of nanofluids", *Renew. Sustain. Energy Rev.*, vol. 15, no. 6, pp. 3271-3277, 2011.
http://dx.doi.org/10.1016/j.rser.2011.04.025

[12] I. Khan, K. Saeed, and I. Khan, "Nanoparticles: Properties, applications and toxicities", *Arab. J. Chem.*, vol. 12, no. 7, pp. 908-931, 2019.
http://dx.doi.org/10.1016/j.arabjc.2017.05.011

[13] X. Ma, Y. Song, Y. Wang, Y. Zhang, J. Xu, S. Yao, and K. Vafai, "Experimental study of boiling heat transfer for a novel type of GNP-Fe_3O_4 hybrid nanofluids blended with different nanoparticles", *Powder Technol.*, vol. 396, pp. 92-112, 2022.
http://dx.doi.org/10.1016/j.powtec.2021.10.029

[14] S. Ottofuelling, F. Von Der Kammer, and T. Hofmann, "Commercial titanium dioxide nanoparticles in both natural and synthetic water: comprehensive multidimensional testing and prediction of aggregation behavior", *Environ. Sci. Technol.*, vol. 45, no. 23, pp. 10045-10052, 2011.
http://dx.doi.org/10.1021/es2023225 PMID: 22013881

[15] Raja Izamshah, Jingsi Wang, S.Y. Chang, M.S. Salleh, R. Izamshah, and J. Wang, "Experimental study of quarry dust and aluminium oxide suspension as cutting fluid for drilling of titanium alloy", *J. Adv. Res. Fluid. Mechan. Therm. Sci.*, vol. 91, no. 2, pp. 145-153, 2022.
http://dx.doi.org/10.37934/arfmts.91.2.145153

[16] H. Alfvén, "Existence of electromagnetic-hydrodynamic waves", *Nature*, vol. 150, no. 3805, pp. 405-406, 1942.
http://dx.doi.org/10.1038/150405d0

[17] A. Raza, R. Ul Haq, S.S. Shah, and M. Alansari, "Existence of dual solution for micro-polar fluid flow with convective boundary layer in the presence of thermal radiation and suction/injection effects", *Int. Commun. Heat Mass Transf.*, vol. 131, p. 105785, 2022.
http://dx.doi.org/10.1016/j.icheatmasstransfer.2021.105785

[18] M.N. Khan, and S. Nadeem, "A comparative study between linear and exponential stretching sheet with double stratification of a rotating Maxwell nanofluid flow", *Surf. Interfaces*, vol. 22, p. 100886, 2021.
http://dx.doi.org/10.1016/j.surfin.2020.100886

[19] M. P. Mkhatshwa, "Overlapping grid spectral collocation methods for nonlinear differential equations modelling fluid flow problems", 2020.

[20] M.P. Mkhatshwa, "Overlapping grid spectral collocation approach for electrical MHD bioconvection Darcy–Forchheimer flow of a Carreau–Yasuda nanoliquid over a periodically accelerating surface", *Heat Transf.,* vol. 51, no. 2, pp. 1468-1500, 2022.
http://dx.doi.org/10.1002/htj.22360

[21] M.P. Mkhatshwa, S.S. Motsa, and P. Sibanda, "Numerical solution of time-dependent Emden-Fowler equations using bivariate spectral collocation method on overlapping grids", *Nonlinear Eng.,* vol. 9, no. 1, pp. 299-318, 2020.
http://dx.doi.org/10.1515/nleng-2020-0017

[22] V.M.S. Motsa, and Z. Makukula, "Simple iteration methods for non-linear differential equations: Theory and Development", *10th Annual Research Workshop on Numerical Methods for Differential Equations,* , 2017

[23] O. Otegbeye, and M.S. Ansari, "A finite difference based simple iteration method for solving boundary layer flow problems", AIP Conference Proceedings, 2022, vol. 2435, no. 1: AIP Publishing LLC, p. 020055.
http://dx.doi.org/10.1063/5.0084396

[24] H.S. Takhar, A.J. Chamkha, and G. Nath, "Unsteady three-dimensional MHD-boundary-layer flow due to the impulsive motion of a stretching surface", *Acta Mech.,* vol. 146, no. 1-2, pp. 59-71, 2001.
http://dx.doi.org/10.1007/BF01178795

[25] H. Xu, S.J. Liao, and I. Pop, "Series solutions of unsteady three-dimensional MHD flow and heat transfer in the boundary layer over an impulsively stretching plate", *Eur. J. Mech. BFluids,* vol. 26, no. 1, pp. 15-27, 2007.
http://dx.doi.org/10.1016/j.euromechflu.2005.12.003

[26] T. Hayat, M. Qasim, and Z. Abbas, "Homotopy solution for the unsteady three-dimensional MHD flow and mass transfer in a porous space", *Commun. Nonlinear Sci. Numer. Simul.,* vol. 15, no. 9, pp. 2375-2387, 2010.
http://dx.doi.org/10.1016/j.cnsns.2009.09.013

[27] D. Halliday, R. Resnick, and J. Walker, *Fundamentals of physics..* John Wiley & Sons, 2013.

[28] M. Sheikholeslami, and D.D. Ganji, "Ferrohydrodynamic and magnetohydrodynamic effects on ferrofluid flow and convective heat transfer", *Energy,* vol. 75, pp. 400-410, 2014.
http://dx.doi.org/10.1016/j.energy.2014.07.089

[29] R. Ahmad, M. Mustafa, T. Hayat, and A. Alsaedi, "Numerical study of MHD nanofluid flow and heat transfer past a bidirectional exponentially stretching sheet", *J. Magn. Magn. Mater.,* vol. 407, pp. 69-74, 2016.
http://dx.doi.org/10.1016/j.jmmm.2016.01.038

[30] J.A. Okello, W.N. Mutuku, and A.O. Oyem, "Analysis of Ethylene Glycol (EG)-based ((Cu-Al_2O_3), (Cu-TiO_2), (TiO_2-Al_2O_3)) hybrid nanofluids for optimal car radiator coolant", *J. Eng. Res. Rep.,* vol. 17, pp. 34-50, 2020.
http://dx.doi.org/10.9734/jerr/2020/v17i217186

[31] A.O. Borode, N.A. Ahmed, P.A. Olubambi, M. Sharifpur, and J.P. Meyer, "Investigation of the thermal conductivity, viscosity, and thermal performance of graphene nanoplatelet-alumina hybrid nanofluid in a differentially heated cavity", *Front. Energy Res.,* vol. 9, p. 737915, 2021.

http://dx.doi.org/10.3389/fenrg.2021.737915

[32] S.M. Hussain, "Dynamics of ethylene glycol-based graphene and molybdenum disulfide hybrid nanofluid over a stretchable surface with slip conditions", *Sci. Rep.,* vol. 12, no. 1, p. 1751, 2022.

http://dx.doi.org/10.1038/s41598-022-05703-z PMID: 35110577

[33] S. Suresh, K.P. Venkitaraj, P. Selvakumar, and M. Chandrasekar, "Synthesis of Al_2O_3–Cu/water hybrid nanofluids using two step method and its thermo physical properties", *Colloids Surf. A Physicochem. Eng. Asp.,* vol. 388, no. 1-3, pp. 41-48, 2011.

http://dx.doi.org/10.1016/j.colsurfa.2011.08.005

[34] J.K. Madhukesh, G.K. Ramesh, G.S. Roopa, B.C. Prasannakumara, N.A. Shah, and S.J. Yook, "3D flow of hybrid nanomaterial through a circular cylinder: Saddle and Nodal Point Aspects", *Mathematics,* vol. 10, no. 7, p. 1185, 2022.

http://dx.doi.org/10.3390/math10071185

[35] S. Rana, M. Nawaz, and M.K. Alaoui, "Three dimensional heat transfer in the Carreau-Yasuda hybrid nanofluid with Hall and ion slip effects", *Phys. Scr.,* vol. 96, no. 12, p. 125215, 2021.

http://dx.doi.org/10.1088/1402-4896/ac2379

[36] M. Revers, "On the approximation of certain functions by interpolating polynomials", *Bull. Aust. Math. Soc.,* vol. 58, no. 3, pp. 505-512, 1998.

http://dx.doi.org/10.1017/S0004972700032494

[37] V.M. Magagula, S.S. Motsa, and P. Sibanda, "A multi-domain bivariate pseudospectral method for evolution equations", *Int. J. Comput. Methods,* vol. 14, no. 4, p. 1750041, 2017.

http://dx.doi.org/10.1142/S0219876217500414

[38] P. Dlamini, S. Motsa, and M. Khumalo, *Higher order compact finite difference schemes for unsteady boundary layer flow problems.,* vol. 2013. Advances in Mathematical Physics, 2013.

CHAPTER 3

Effects of Activation Energy and Thermo-Convection of Nanofluid Flow

Sewli Chatterjee[1,*]

[1] *Department of Mathematics, Turku Hansda Lapsa Hemram Mahavidyalay, (Under Burdwan University), Mallarpur, Madian, West Bengal, India*

Abstract: Nonlinear mixed convection in magnetohydrodynamic thermo-convection nanofluid flow is addressed. Flow is induced by a stretched surface. Chemical reaction effects on the activation energy of the species concentration of Arrhenius are considered. The nonlinear governing equations are converted to a dimensionless ordinary system through adequate transformations. The solutions of nonlinear dimensionless expressions are computed by employing the spectral quasilinearization method. A consequence of various parameters involved such as velocity, concentration, and thermal field has been reported graphically. Concentration has the reverse trend for both random motion and thermophoresis variables.

Keywords: Arrhenius energy, Activation energy, SQLM (Spectral Quasilinearization method), Thermos-convection.

INTRODUCTION

Current uses of the borderline fluid stream over a continuous stretching sheet include the production of paper, glass fibre, polymer extrusion, wire and plastic film manufacturing, and many more industries where the fluids' freeze and heat properties are controlled, which grab the eyeballs of the many researchers. From the time it was introduced by Lewig Prandtl in 1904, various studies by different researchers have been carried out [1-5]. Boundary layer fluid flows are commonly used in a variety of engineering applications, including geophysical drifts, groundwater hydrology, petroleum lochs, electronic system refrigeration, porcelain processes, categorization processes, current padding, groundwater smog, and compound catalytic reactors. Various researchers like [6-7] examine the non-Darcy porous medium with micro-convection nanofluids. Various researchers [8-12] discussed classifying the motile microorganism with viscous dissipation and

[*]**Corresponding author Sewli Chatterjee:** Department of Mathematics, Turku Hansda Lapsa Hemram Mahavidyalay, (Under Burdwan University), Mallarpur, Madian, West Bengal 731216, India; E-mail: sewlichatterjee81@gmail.com

Sabyasachi Mondal (Ed.)

activation energy. The effect of activation energy and thermo-convective nanofluid flow is being addressed by many scholars [13-16]. Investigation of the Nanofluid fluid depends on different base fluids that are important at present days for industrial as well as for engineering sectors. Nanofluids are made of nanosized oxide or metal oxides mixed in a proper base fluid like water, or ethylene glycol. These special kinds of fluids are significant to enhance the heat and mass transfer rate. In biomedical studies, the use of nanofluids is also extremely significant.

The chemical reaction is subject to Arrhenius' activation energy with magnetic parameters including heat generation as a new concept. Velocity and micro-convective boundary conditions are considered here. Brownian motion is considered in this study as a parameter having a great influence on nanofluids with a magnetic effect. The behaviour of micro-convective nanofluid flow is gaining interest nowadays. Taxis and kinetics are two kinds of processes by which micro-convection can happene. They respond to stimuli from many sources, like light, chemical reactions, and temperature. Houseflies are attracted by the light while mosquitoes move away from light, this behaviour of insects is known as positive and negative taxis, respectively.

Our aim through this research is to examine the mathematical model containing two-dimensional MHD steady stream, mass, and heat transport incompressible micro-convective nanofluids with Arrhenius activation energy. A suitable similarity transformation has been used to non–dimensionalize the system of non-dimensional ordinary through adequate transformations. The converted nonlinear coupled ODEs have been solved by operating the Spectral Quasi-linearization technique numerically. The impacts of key parameters have been investigated diagrammatically.

MATHEMATICAL MODEL

Here a two-dimensional magnetohydrodynamics thermo-convection nanofluid flow with Arrhenius activation energy has been taken into consideration. The mathematical form is $u_w(x) = cx$, where is $c > 0$ (Fig. **1**).

$$\frac{\partial u}{\partial x} + \frac{\partial v}{\partial y} = 0 \tag{1}$$

$$u\frac{\partial u}{\partial x} + v\frac{\partial u}{\partial y} = \vartheta\frac{\partial^2 u}{\partial y^2} - \frac{\sigma B_0^2}{\rho_f(1+N^2)}u + (1 - C_\infty)\rho_f g\beta_T(T - T_\infty) - \frac{(\rho_p - \rho_f)}{\rho_f}g\beta_C(C - C_\infty) \tag{2}$$

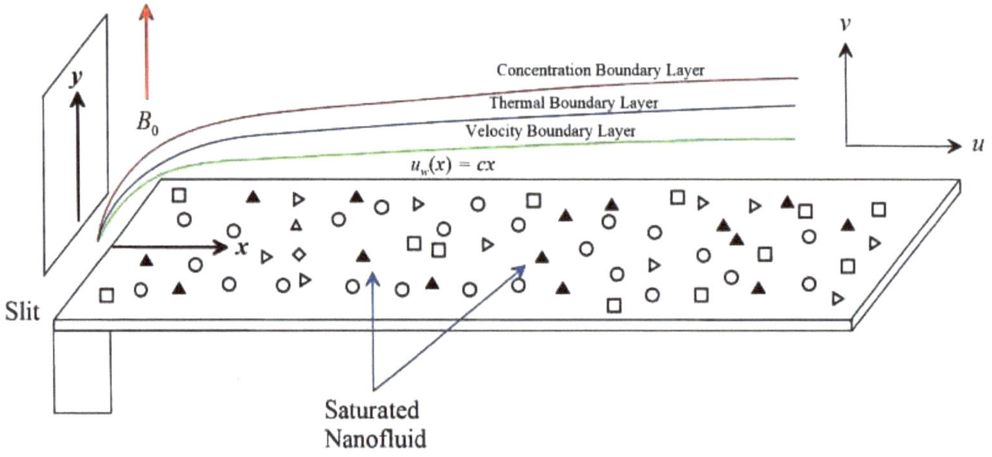

Fig. (1). Physical model.

The governing equations of fluid flow for this present model are defined as follows:

$$u\frac{\partial T}{\partial x} + v\frac{\partial T}{\partial y} = \left(\alpha + \frac{16\sigma^* T_\infty^3}{3\rho_f C_p}\right)\frac{\partial^2 T}{\partial y^2} + \frac{\mu}{\left(\rho C_p\right)_f}\left(\frac{\partial u}{\partial y}\right)^2$$
$$+\tau\left\{D_B\frac{\partial T}{\partial y}\frac{\partial C}{\partial y} + \frac{D_T}{T_\infty}\left(\frac{\partial T}{\partial y}\right)^2\right\}$$
$$+Q(T - T_\infty). \quad (3)$$

$$u\frac{\partial C}{\partial x} + v\frac{\partial C}{\partial y} = D_B\frac{\partial^2 C}{\partial y^2} - k_r^2\left(\frac{T}{T_\infty}\right)^n \exp\left(\frac{-E_a}{kT}\right)(C - C_\infty) + \frac{D_T}{T_\infty}\frac{\partial^2 T}{\partial y^2}. \quad (4)$$

The suitable boundary conditions are considered from Eqs. (1)-(4) as follows:

$$u = cx\,, \quad v = 0, \quad -k_f\frac{\partial T}{\partial y} = h_f(T - T_\infty), \; D_B\frac{\partial C}{\partial y} + \frac{D_T}{T_\infty}\frac{\partial T}{\partial y},=0, \quad \text{at } y = 0\,.$$
$$u = 0\,, \quad T = T_\infty\,, C = C_\infty \text{ at } y = \infty\,. \quad (5)$$

Eqs. (1) and (2) represent the momentum along $x - axis$ and $y - axis$, respectively. σ and ρ_f , ρ_p , are the electrical conductivity, and density of fluid and nanoparticles, respectively. γ , and σ^* are the Stephan-Boltzman constant and microorganism expansion coefficient, respectively.

The Arrhenius function is given by,

$T = \left(\frac{T}{T_\infty}\right)^n e^{\left(\frac{-E_a}{kT}\right)}$, E_a is the activation energy, n is constant.

To covert the partial differential Eqs. (1) to (4) into non-linear ODEs, here we use the following similarity transformation:

$$\psi = \sqrt{c\vartheta}xf(\eta)\,, = y\sqrt{\frac{c}{\vartheta}}\,, \theta(\eta) = \frac{T-T_\infty}{T_w-T_\infty}\,,\ \emptyset(\eta) = \frac{C-C_\infty}{C_w-C_\infty}, \quad \textbf{(6)}$$

By using the above transformation (6), we get:

$$f''' + ff'' - f'^2 - \frac{M}{(1+N^2)}f' + \frac{Gr}{Re^2}(\theta - Nr\emptyset) = 0. \quad \textbf{(7)}$$

$$\left(1 + \frac{4}{3}Rd\right)\theta'' + Pr\left(f\theta' + Ec.f''^2 + Nb.\theta'\emptyset' + Nt\theta'^2\right) + Pr(\delta\theta) = 0. \quad \textbf{(8)}$$

$$\emptyset'' + Sc\left\{f\emptyset' - \lambda_m^2(1 + \varepsilon\theta)^n exp\left(\frac{-E}{1+\varepsilon\theta}\right)\emptyset\right\} + \frac{Nt}{Nb}\theta'' = 0. \quad \textbf{(9)}$$

Boundary conditions also have the non-dimensional form as follows:

$$f'(0) = 1,\ \ f'(\infty) = 0, f(0) = 0.$$

$$\theta'(0) = -Bi_t\left(1 + \theta(0)\right), \theta(\infty) = 0.$$

$$Nb\emptyset'(0) + Nt\theta'(0) = 0,\ \ \emptyset(\infty) = 0. \quad \textbf{(12)}$$

The parameters involved in the above-transformed equations are as follows:

$M = \frac{\sigma B_0^2}{c\rho_f}$ is the magnetic field, $Gr = \frac{(1-C_\infty)\rho_f g\beta_T(T-T_\infty)x^3}{\vartheta^2}$ is the Grashof number, $Re = \frac{cx^2}{\vartheta}$ is the Reynold number, $Nr = \frac{(\rho_p-\rho_f)\beta_C\Delta C}{(1-C_\infty)\rho_f{}^2\beta_T\Delta T}$ is the Buoyancy parameter, $\Pr = \frac{\vartheta}{\alpha}$ Prandtl number, $Ec = \frac{(cx)^2}{c_p\Delta T}$ the Eckert number, $Nb = \frac{\tau D_B\Delta C}{\vartheta}$ the Brownian motion parameter, $Nt = \frac{\tau D_T\Delta T}{\vartheta T_\infty}$ the thermophoresis parameter, $Sc = \frac{\vartheta}{D_B}$ the Schmidt number, $\lambda_m^2 = \frac{kr^2}{c}$ is the chemical reaction rate, $Rd = \frac{4\sigma^* T_\infty^3}{\rho_f\,\alpha\,C_p}$ is the thermal radiation parameter, $\varepsilon = \frac{\Delta T}{T_\infty}$ temperature difference parameter, $B_i = \frac{h_f}{k_f}\frac{x}{\sqrt{Re}}$ the Biot number, where h_f denotes the convective heat transfer coefficient.

To obtain some physical attention which will be considerable for both bioconvection nanofluids flow,

the coefficient of Skin friction is defined as,

$$C_f = \frac{\tau_w}{\frac{1}{2}\rho_f U^2} = \frac{-2\, f''(0)}{\sqrt{Re}}. \tag{13}$$

The local Nusselt number is defined as:

$$N_{u_x} = \frac{x q_w}{k_f(T_w - T_\infty)} = -\left(1 + \frac{4}{3}Rd\right)\sqrt{Re}\,\theta'(0), \tag{14}$$

The local Sherwood number is defined as:

$$S_{h_x} = \frac{x q_c}{D_B(C_w - C_\infty)} = \sqrt{Re}\,\emptyset'(0), \tag{15}$$

Table 1. Comparison of the present results spectral quasi linearization methods.

		Arpita *et al.* [17]		Present Result	
M	δ	$-f''(0)$	$-\left(1+\frac{4}{3}Rd\right)\theta'(0)$	$-f''(0)$	$-\left(1+\frac{4}{3}Rd\right)\theta'(0)$
0.1	0.10	1.03811346	0.11112079	1.03811455	0.11112800
0.2		1.07555091	0.10364227	1.07555111	0.10364300
0.3		1.11170450	0.09592125	1.11170690	0.09592345
0.5		1.18061230	0.07960153	1.18061440	0.07960455
	0.10	1.04831061	0.08975695	1.04831112	0.08975789
	0.15	1.04812974	0.08757069	1.04812094	0.08757700
	0.20	1.04790248	0.08483995	1.04790233	0.08483100
	0.25	1.04760777	0.08132172	1.04760779	0.08132232

RESULT AND DISCUSSION

The validation of the present results with published works that are in excellent agreement has been obtained (Table **1**). Fig. (**2a** and **2b**) show magnetic field effect on Bi-direction velocity profile and thermal profile. Physically magnetic fields produce the flow resisting Lorentz force. This force reduces the velocity profile (a). But the opposite behaviour can be shown for the temperature.

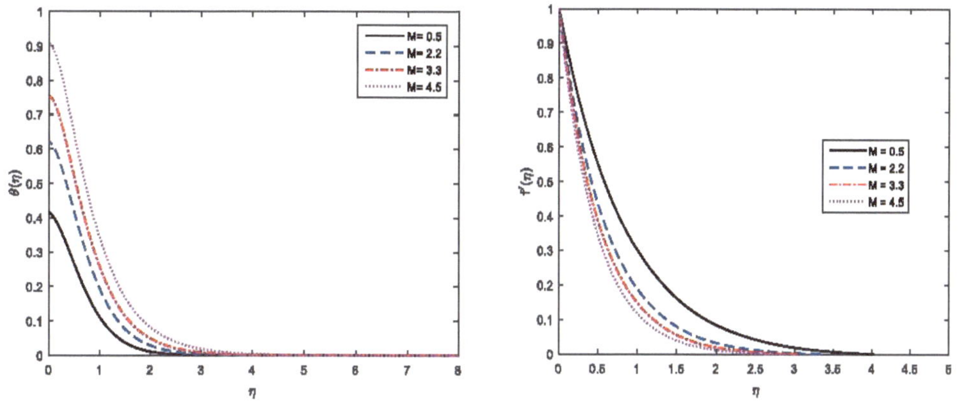

(a) (b)

Fig. (2). (**a**) Impact of Magnetic parameter on momentum, (**b**) Impact of Magnetic parameter on temperature.

In Fig. (**3a** and **3b**), the effects of thermal radiation on temperature and concentration are presented. According to the following figure, it is clear that with an increment in thermal radiation, the temperature profile also increases for comparative values of Prandtl number. The body temperature increases at a non-linear rate with radiation enhancement. The opposite nature of nanofluid concentration can be shown, for the thermal profile, it decreases, whereas, at the starting time, it exhibits a decrease in nature but as far as it goes, this profile increases.

In Figs. (**4a** and **4b**), the impact of thermophoresis on temperature, and nanofluid concentration is presented. In the first Fig. (a) it is very clear that with an increment of thermophoresis, the nanofluid concentration enhanced. Fig. (b) shows the similar behaviour with the thermal boundary layer and solutal boundary layer.

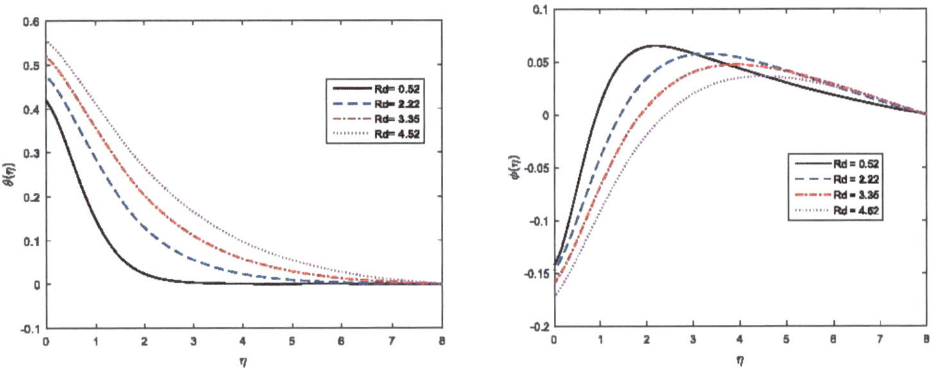

(a) (b)

Fig. (3). Effects of Radiation *Rd* on temperature (**a**) and nanofluid concentration (**b**).

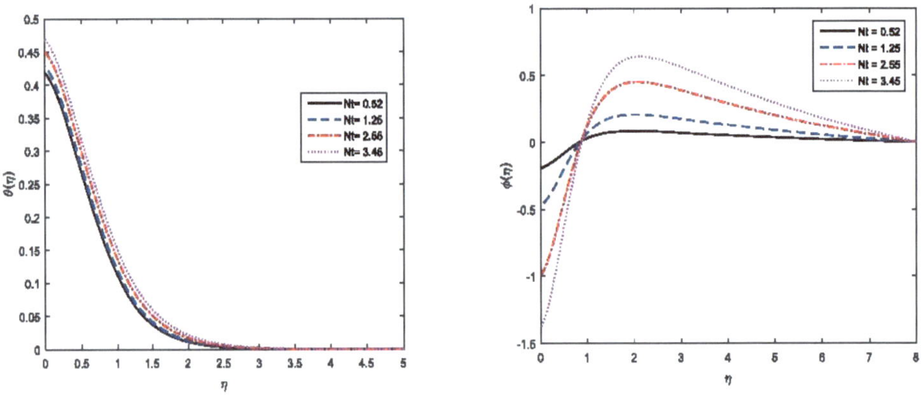

(a) (b)

Fig. (4). Effect of thermophoresis on temperature (**a**) and concentration (b).

In Fig. (**5**) the impact of the Brownian motion parameter on concentration is presented. In this figure, it is very clear that an increment of Brownian motion, the concentration enhanced with the solutal boundary layer.

Physically Schmidt number is the ratio of viscosity to mass diffusivity. Body temperature increases with the increment of Schmidt number for different values of heat generation parameter (Fig. **6a**), but with to a reduction in mass diffusivity, the opposite behaviour for Micro-organism concentration was observed. It diminishes for larger values of Schmidt number (Fig. **6b**).

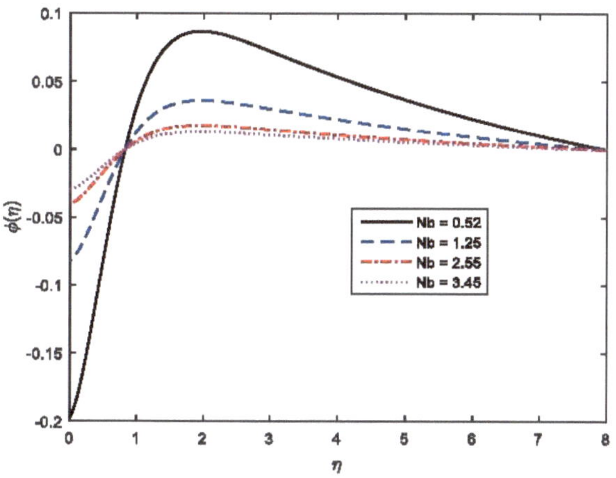

Fig. (5). Impact of Brownian motion parameter on solutal concentration.

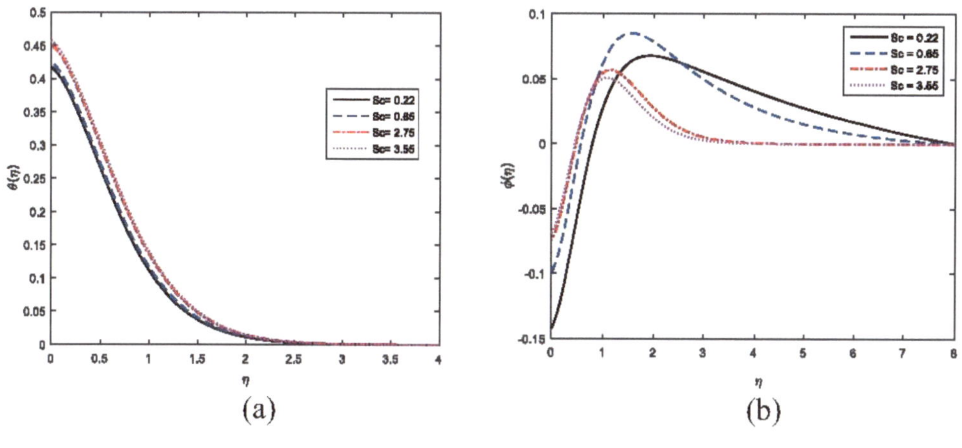

Fig. (6). Effect of Schmidt number on (**a**) temperature and (**b**) concentration fields.

The sequel of the Eckert number on various boundary layers has been shown in Figs. (**7a** and **7b**). It is very clear that this is a key parameter for the current research, each effect is described very transparently. Temperature increased due to the increment in Eckert number, a similar nature can be seen for each concentration also.

Figs. (**8a** and **8b**) present the sequel of the heat generation parameter. It is a very interesting fact that temperature profiles decrease with the enhancement of heat generation parameter. However, the concentration deteriorates far from the boundary layer.

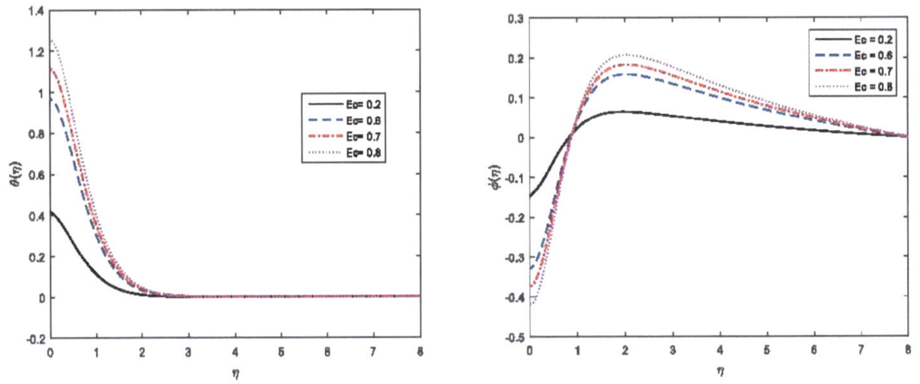

(a) (b)

Fig. (7). (**a**) Impact of Eckert number on (**a**) temperature, and (**b**) solutal concentration effects.

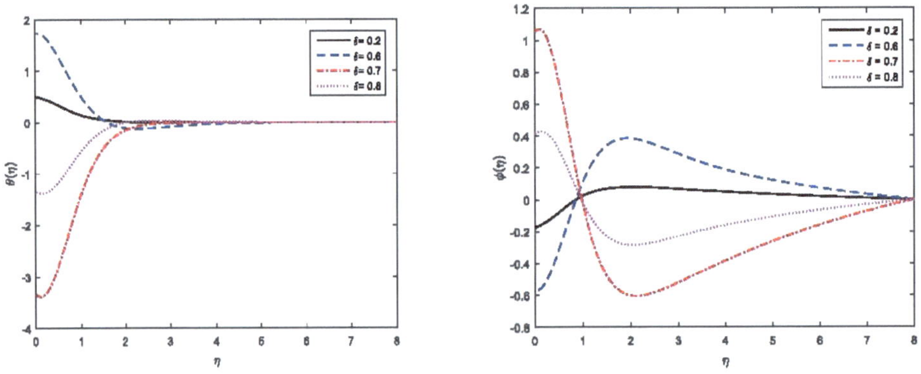

(a) (b)

Fig. (8). (**a**) Impact of heat generation parameter on (**a**) temperature and (**b**) concentration profiles.

CONCLUSION

We tried to develop a mathematical model of micro-mixed convection in nanofluid. This investigation comprises the internal heat generation and activation energy. In this paper, we found various key parameters in nature throughout the motion. Some of the leading observations are listed below:

- Magnetic parameter decreases the concentration but increases the temperature. We examined it by taking several values of the Prandtl number and made this significant conclusion.
- The radiation parameter enhances the temperature of the system.
- Enhancement of Schmidt's number decreases the microbial concentration but enhances the temperature.

REFERENCES

[1] E. Magyari, and B. Keller, "Heat and mass transfer in the boundary layers on an exponentially stretching continuous surface", *J. Phys. D Appl. Phys.*, vol. 32, no. 5, pp. 577-585, 1999. http://dx.doi.org/10.1088/0022-3727/32/5/012

[2] T. Orikasa, S. Koide, S. Okamoto, C. Togashi, T. Komoda, S. Hatanaka, Y. Muramatsu, M. Thammawong, T. Shiina, and A. Tagawa, "Temperature dependency of quality change during far-infrared drying of komatsuna leaves", *Acta Hortic.*, no. 1091, pp. 319-325, 2015. http://dx.doi.org/10.17660/ActaHortic.2015.1091.40

[3] P. Sibanda, M. Almakki, Z. Mburu, and H. Mondal, "Entropy optimization in MHD nanofluid flow over an exponential stretching sheet", *Appl. Sci.*, vol. 12, p. 10809, 2022. http://dx.doi.org/10.3390/app122110809

[4] M. Dhlamini, H. Mondal, P. Sibanda, and S. Motsa, "Numerical analysis of couple stress nanofluid in temperature dependent viscosity and thermal conductivity", *Int. J. Appl. Comput. Math.*, vol. 7, no. 2, p. 48, 2021. http://dx.doi.org/10.1007/s40819-021-00983-x

[5] M. Almakki, H. Mondal, P.K. Kameswaran, and P. Sibanda, "Entropy generation of double diffusive convection magnetic nanofluid in the stagnation region with rotating sphere in presence of viscous dissipation", *J. Nanofluids,* vol. 11, pp. 360-372, 2022. http://dx.doi.org/10.1166/jon.2022.1847

[6] I. Nasser, and H. Duwairi, "Thermal dispersion effects on convection heat transfer in porous media with viscous dissipation", *Int. J. Heat . Technol.*, vol. 34, no. 2, pp. 207-212, 2016. http://dx.doi.org/10.18280/ijht.340208

[7] R. Kairi, "Viscosity and dispersion effects on natural convection from a vertical cone in a nonnewtonian fluid saturated porous medium", *Therm. Sci.*, vol. 15, no. 2, suppl. 2, pp. 307-316, 2011. http://dx.doi.org/10.2298/TSCI110614124K

[8] K. Sruthila Gopalakrishnan, I.S. Oyelakin, S. Mondal, and R.P. Sharma, "Impact of Joule heating and nonlinear thermal radiation on the flow of Casson nanofluid with entropy generation", *Int. J. Amb. Energy,* vol. 43, no. 1, pp. 5687-5710, 2022. http://dx.doi.org/10.1080/01430750.2021.1973559

[9] S. Siddiqa, Gul-e-Hina, N. Begum, S. Saleem, M.A. Hossain, and R.S. Reddy Gorla, "Numerical solutions of nanofluid bioconvection due to gyrotactic microorganisms along a vertical wavy cone", *Int. J. Heat Mass Transf.*, vol. 101, pp. 608-613, 2016. http://dx.doi.org/10.1016/j.ijheatmasstransfer.2016.05.076

[10] Z. Iqbal, Z. Mehmood, and E.N. Maraj, "Oblique transport of gyrotactic microorganisms and bioconvection nanoparticles with convective mass flux", *Physica E,* vol. 88, pp. 265-271, 2017.

http://dx.doi.org/10.1016/j.physe.2017.01.011

[11] W.A. Khan, M.J. Uddin, and A.I.M. Ismail, "Free convection of non-Newtonian nanofluids in porous media with gyrotactic microorganisms", *Transp. Porous Media,* vol. 97, no. 2, pp. 241-252, 2013.

http://dx.doi.org/10.1007/s11242-012-0120-z

[12] A.V. Kuznetsov, "Nanofluid bioconvection: interaction of microorganisms oxytactic upswimming, nanoparticle distribution, and heating/cooling from below", *Theor. Comput. Fluid Dyn.,* vol. 26, no. 1-4, pp. 291-310, 2012.

http://dx.doi.org/10.1007/s00162-011-0230-1

[13] H.S. Takhar, A.J. Chamkha, and G. Nath, "MHD flow over a moving plate in a rotating fluid with magnetic field, Hall currents and free stream velocity", *Int. J. Eng. Sci.,* vol. 40, no. 13, pp. 1511-1527, 2002.

http://dx.doi.org/10.1016/S0020-7225(02)00016-2

[14] S. Noreen, and M. Qasim, "Influence of hall current and viscous dissipation on pressure driven flow of pseudoplastic fluid with heat generation: a mathematical study", *PLoS One,* vol. 10, no. 6, p. e0129588, 2015.

http://dx.doi.org/10.1371/journal.pone.0129588 PMID: 26083027

[15] H.G. Lee, and J. Kim, "Numerical investigation of falling bacterial plumes caused by bioconvection in a three-dimensional chamber", *Eur. J. Mech. BFluids,* vol. 52, pp. 120-130, 2015.

http://dx.doi.org/10.1016/j.euromechflu.2015.03.002

[16] R. Ghosh, T.M. Agbaje, S. Mondal, and S. Shaw, "Bio-convective viscoelastic Casson nanofluid flow over a stretching sheet in the presence of induced magnetic field with Cattaneo–Christov double diffusion", *Int. J. Biomath.,* vol. 15, no. 3, p. 2150099, 2022.

http://dx.doi.org/10.1142/S1793524521500996

[17] A. Mandal, H. Mondal, and R. Tripathi, "Activity of motile microorganism in bioconvective nanofluid flow with Arrhenius activation energy", *J. Therm. Anal. Calorim.,* vol. 148, no. 17, pp. 9113-9130, 2023.

http://dx.doi.org/10.1007/s10973-023-12295-x

CHAPTER 4

Study of Oblique Nanofluid Flow Past a Stretching Sheet

Vikas Poply[1,*], **Shilpa Taneja**[2] and **Parveen Kumar**[3]

[1] *Department of Mathematics, KLP College Rewari, Haryana, India*

[2] *Department of Physics, N.B.G.S.M. College, Sohna, Gurugram, Haryana, India*

[3] *Department of Physics, Govt. P.G. Nehru College, Jhajjar, Haryana, India*

Abstract: The analysis of impinging oblique nano-fluid past a stretching sheet is done in the current communication. The nanoparticles' behaviour in fluids for convective heat transfer has been extensively studied. We assume that the fluid and the stretching velocity change linearly. In this paper, we used a cutting-edge Genetic Algorithm technique to better understand how a nanofluid's thermal characteristics and heat transport vary over time. Additionally, the impact of a change in the fluid's impinging angle is investigated. The temperature and concentration curves, skin friction coefficient, and Nusselt number have all been observed and graphically displayed.

Keywords: Genetic algorithm, Heat transfer, Nanofluids.

INTRODUCTION

The traditional transport of heat in nanofluids has recently gained attention as a coeval topic. Choi was the first to use the term "nanofluid" [1]. A base fluid and nanoparticles are combined to form a fluid called a nanofluid. With their underlying fluids, they have extremely high thermal conductivity. Nanoparticles have a diameter that falls between 1 and 100 nm.

Typically, the metals (Al, Cu), oxides, carbides, nitrites, and carbon nanotubes make up the nanoparticles employed in nanofluids (allotrope of carbon nonmetal). When combined with other substrates, nanoparticles have the inherent ability to improve thermal conductivity. Base fluids include conductive liquids like water, oil, ethylene glycol, *etc*.

[*]**Corresponding author Vikas Poply:** Department of Mathematics, KLP College Rewari, Haryana, India; E-mail: vikaspoply@gmail.com

The conventional heat transfer fluids' thermal conductivity is insufficient to meet modern industries' demands for cooling rates. Considered as cutting-edge technology to ensure nuclear safety is a nanofluid coolant with improved thermal performance. In order to guarantee a functional improvement in heat transmission, nanoparticles commonly possess up to a modest 5% volume fraction of nanoparticles. An experimental investigation disclosed that a slight volumetric percentage of nanoparticles significantly increased the thermal conductivity by 10–50% with an impressive mastery of the fluid.

Buongiorno [2] provides a thorough overview of convective transference in nanofluids. He discovered that the slip velocity can be expected to add up to the nanoparticle's velocity (relative velocity). Following his reasoning, he concluded that, only Brownian motion and thermophoresis are momenta. Based on the mechanics of the relative velocities of the nanoparticle and base fluid, Buongiorno suggested a new model.

Various processes involve pushing when an incompressible viscous fluid flows across a stretched surface. Modern academics have been drawn to the current topic because of its amazing applications. Crane [3] investigated the effects of applying uniform stress to a 2-D continuous flow and concluded that the stretching of an elastic sheet changes linearly from a constant point.

The stagnation point flow approaching the extending surface was examined by Gupta and Mahapatra [4]. Singh *et al.* [5-8] investigated how magnetic parameters and radiations affected several aspects of the stretched sheet. Recently, the study of magnetohydrodynamic flow is of utmost importance. The deformation of a fluid-containing vessel's wall causes magnetohydrodynamic flow to occur. A unique substance with both liquid and magnetic qualities is known as magnetic nanofluid. By changing the magnetic field, it is possible to regulate several physical characteristics. Hamad [9] investigated the convective flow of nanofluid *via* a semi-infinite vertical sheet. In a nanofluid, Bachok *et al.* [10, 11] investigated the boundary layer flow toward a stretching sheet.

Genetic algorithm, the main paradigm of evolutionary computation, is an optimization technique. It relies on Darwin's theory of the "survival of the fittest" that drives biological evolution. John Holland [12] was the inventor of the genetic algorithm who described the technique in his book "Adaptation in Natural and Artificial Systems". Lots of researchers and scientists since then have contributed to this area with varying degrees of success. McCall [13] has explained that the genetic algorithm is an active and flourishing area that has the potential to reach the zenith. It can explain even the complex phenomenon that was out of reach for the earlier theories. Chaiyartna and Zalzala [14] have reviewed the developments in this area. A technique for resolving optimization issues based on the idea of natural selection is the genetic algorithm. The genetic algorithm employs three important processes, first, the selection where the individuals are randomly selected from the available parents, second the crossover where the selected parents combine to form the next generation and the third mutation in which random changes occur at the parent level to form children. The population formed after successive iterations evolves towards an optimized solution. It has an edge over the standard optimization algorithms in various linear and non-linear physical problems, such as heat and fluid flow because of its highly modular. In recent days, these flows are either studied by finite difference method with shooting technique or other optimization techniques. These obsolete methods falter at the front that they require an initial guess for the function to be optimized. Soft computing algorithms are the best and most reliable alternative for getting the optimum design in fluid dynamics.

FORMULATION OF THE PROBLEM

A stretching sheet having invariable temperature T_w is the physical issue being considered here on which the viscous nanofluid is incident obliquely means at a certain angle of incidence (Fig. **1**). The nanofluid has characteristics of incompressibility, steady flow in two dimensions in the upper half plane and good electrical conductivity. The stretching surface is positioned along the x-axis and in the plane y = 0. A magnetic field of transverse nature is applied on this nanofluid which possesses the non-uniform velocity of the form of u_w = Ax where A is a positive constant having dimension (time) −1.

The Joule heating effect, the magnetic field that can be created in the due course of flow, and the dissipation caused by the fluid's viscous nature were some of the aspects that we overlooked while examining this physical issue. The various conservation equations strictly adhering to above stated assumptions may be written as under-

$$\frac{\partial u}{\partial x}+\frac{\partial v}{\partial y}=0 \tag{1}$$

$$u\frac{\partial u}{\partial x}+v\frac{\partial u}{\partial y}=-\frac{1}{\rho_b}\frac{\partial p}{\partial x}+\upsilon\left(\frac{\partial^2 u}{\partial x^2}+\frac{\partial^2 u}{\partial y^2}\right)-\sigma\frac{u}{\rho}B^2 \tag{2}$$

$$u\frac{\partial v}{\partial x}+v\frac{\partial v}{\partial y}=-\frac{1}{\rho_b}\frac{\partial p}{\partial y}+\upsilon\left(\frac{\partial^2 v}{\partial x^2}+\frac{\partial^2 v}{\partial y^2}\right) \tag{3}$$

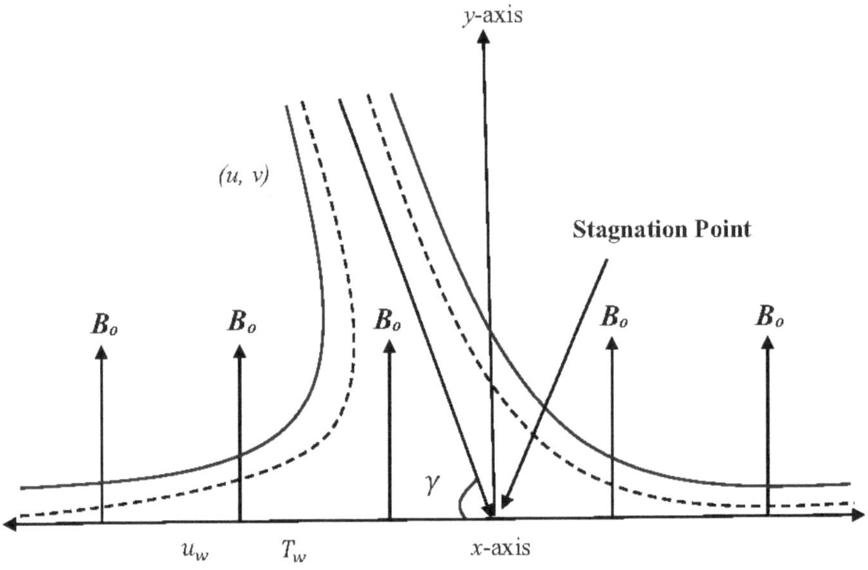

Fig. (1). Schematic diagram.

$$u\frac{\partial T}{\partial x}+v\frac{\partial T}{\partial y}=\alpha\left(\frac{\partial^2 T}{\partial x^2}+\frac{\partial^2 T}{\partial y^2}\right)+\tau\left\{D_b\left(\frac{\partial C}{\partial x}\frac{\partial T}{\partial x}+\frac{\partial C}{\partial y}\frac{\partial T}{\partial y}\right)+\frac{D_t}{T_\infty}\left[\left(\frac{\partial T}{\partial x}\right)^2+\left(\frac{\partial T}{\partial y}\right)^2\right]\right\} \quad (4)$$

$$u\frac{\partial C}{\partial x}+v\frac{\partial C}{\partial y}=D_b\left(\frac{\partial^2 C}{\partial x^2}+\frac{\partial^2 C}{\partial y^2}\right)+\frac{D_t}{T_\infty}\left(\frac{\partial^2 T}{\partial x^2}+\frac{\partial^2 T}{\partial y^2}\right), \quad (5)$$

where the symbols in the above equations are denoted as—

$u,$ and v represent the velocity component in the x and y direction, respectively. T is the temperature, $v, \alpha, D_t, D_{b,}, C$ represents kinematic viscosity, thermal diffusivity, thermophoresis diffusion coefficient, Brownian diffusion coefficient, volumetric volume coefficient, σ is the electrical conductivity, and τ is the ratio between heat capacity of nanoparticle material and heat capacity of the fluid.

The boundary restrictions are:

$$u=u_w(x)=Ax\,,v=0\,,T=T_w\,,C=C_w\quad \text{at}\ y=0$$

$$u=ax\sin\gamma+by\cos\gamma\,,\ v=-ay\sin\gamma\,,T=T_\infty\,,C=C_\infty\ \text{as} y\to\infty \quad (6)$$

where a, b, and A are constants and T_∞ and C_∞ are the temperature and concentration of the fluid away from the surface and γ is the angle parameter.

The stream function $\psi(x,y)$ is defined as $u = \frac{\partial\psi}{\partial y}$ *and* $v = -\frac{\partial\psi}{\partial x}$ and the similarity variables are defined as $\eta = \sqrt{\frac{A}{\nu}}\, y,\ \xi = \sqrt{\frac{A}{\nu}}\, x$.

Using these boundary conditions, the equation becomes,

$$\psi = 0,\ \frac{\partial\psi}{\partial\eta} = \xi, \theta(0) = 1,\ \phi(0) = 1 \text{ at } \eta = 0$$

$$\psi = \lambda\xi\eta\sin\gamma + \frac{1}{2}k\eta^2\cos\gamma\ \theta \to 0\,,\ \phi \to 0 \text{ as } \eta \to \infty \tag{7}$$

$\lambda = a/A$ is the stretching sheet parameter and $k = \frac{b}{A}$ is constant.

We need the result of the form of $\psi = \xi f(\eta) + g(\eta)$ where the function $f(\eta)$ and $g(\eta)$ denote the normal and tangential component flow. So the velocity can be written as:

$$u(\xi,\eta) = \frac{\partial\psi}{\partial y} = \xi f'(\eta)\frac{\partial\eta}{\partial y} + g'(\eta)\frac{\partial\eta}{\partial y} \tag{8}$$

$$u(\xi,\eta) = \sqrt{\frac{A}{\nu}}(\xi f'(\eta) + g'(\eta)) \text{ and } v(\xi,\eta) = -\frac{\partial\psi}{\partial x} = -\sqrt{\frac{A}{\nu}}\, f(\eta) \tag{9}$$

Continuity Eq. (1) is satisfied by $u(\xi,\eta)$ *and* $v(\xi,\eta)$ defined as above and using it in Eq. (7), we get:

$$\begin{aligned} f'(\eta)\big(\xi f''(\eta) + g''(\eta)\big) - \big(\xi f'(\eta) + g'(\eta)\big)F''(\eta) - M\big(\xi f''^{(\eta)} + g''^{(\eta)}\big) \\ - f'(\eta)\big(\xi f''(\eta) + g''(\eta)\big) + f(\eta)\big(\xi f'''(\eta) + g'''(\eta)\big) \\ + \big(\xi f''''(\eta) + g''''(\eta)\big) = 0 \end{aligned}$$

Comparing the coefficient of $\xi$and ξ^0, we get:

$$f'''(\eta) + f(\eta)f''(\eta) - \big(f'(\eta)\big)^2 - Mf'(\eta) + C_1 = 0 \tag{10}$$

$$g'''(\eta) + f(\eta)g''(\eta) - f'(\eta)g'(\eta) - Mg'(\eta) + C_2 = 0 \quad \textbf{(11)}$$

where C_1 and C_2 are constants and determined by boundary restrictions.

$$f(0) = 0\,, f'(0) = 1\,, g(0) = 0\,, \; g'(0) = 0$$

$$f'(\infty) = \lambda \sin\gamma \;, g''(\infty) = k\cos\gamma \quad \textbf{(12)}$$

Using the above, the governing equations reduce to –

$$f'''(\eta) + f(\eta)f''(\eta) - \left(f'(\eta)\right)^2 - M(f'(\eta) - \lambda\sin\gamma) + (\lambda\, sin\,\gamma)^2 = 0 \quad \textbf{(13)}$$

$$g'''(\eta) + f(\eta)g''(\eta) - f'(\eta)g'(\eta) - k\cos\gamma - M(g'(\eta) - k\eta\, cos\,\gamma) = 0 \quad \textbf{(14)}$$

Similarly, equations (4) and (5) can be modified with the help of the dimensionless temperature and the dimensionless concentration.

$$\theta = \frac{T - T_\infty}{T_w - T_\infty} \qquad and \qquad \phi = \frac{C - C_\infty}{C_w - C_\infty}$$

And boundary restrictions are:

$$\theta(0) = 1\,, \phi(0) = 1\,, \theta(\infty) = 0 \;\; \text{and} \;\; \phi(\infty) = 0$$

$$\theta''(\eta) + Pr\, f(\eta)\theta'(\eta) + PrN_b\phi'(\eta)\theta'(\eta) + PrN_b\left(\theta'(\eta)\right)^2 = 0 \quad \textbf{(15)}$$

$$\phi''(\eta) + Lef(\eta)\phi'(\eta) + \left(\frac{N_t}{N_b}\right)\theta''(\eta) = 0 \quad \textbf{(16)}$$

where prime represents differentiation w.r.t. η and $M = \frac{\sigma B^2}{A\rho_b}, Pr = \frac{v}{\alpha}, Le = \frac{v}{D_b}, N_b = \frac{(A\rho)_p D_b(\phi_w - \phi_\infty)}{(A\rho)_p v}, N_t = \frac{(A\rho)_p D_t(T_w - T_\infty)}{(A\rho)_b T_\infty v}$ symbolizes magnetic parameter (Hartmann number), Prandtl number, Lewis number, Brownian motion parameter, and thermophoresis parameter.

The quantities of real-world attention in the investigation are the Nusselt number Nu and the Sherwood number Sh, defined as -

$$Nu = \frac{x\, q_w}{K\,(T_w - T_\infty)} \qquad ; \qquad Sh = \frac{x\, q_m}{D_b(C_w - C_\infty)}$$

where q_w and $q_m$are the wall heat and mass fluxes, respectively.

Using similarity variables, we obtain,

$$Re_x^{-1/2}\,Nu = -\theta'(0); \qquad Re_x^{-1/2}\,Sh = -\phi'(0)$$

Where $Re_x^{-1/2} = u_w(x)x/\nu$ denotes the Reynolds number. Other quantities of skin friction, may be understood in terms of shear stress τ_w

$\tau_w = \mu(\partial u/\partial x + \partial u/\partial y)|_{y=0}$. Therefore $\tau_w = \mu(\xi f''(0) + g''(0))$.

RESULTS AND DISCUSSION

The RKF method with genetic algorithm has been used to solve equations with boundary restrictions for values of M, λ, N_b, $N_t$and Le taking step size 0.001.

Table 1. Comparison of $-\theta'(0)$ for *Pr*.

Pr	Khan and Pop [15]	Present Result
0.70	0.4539	0.454
2	0.9114	0.9113
7	1.8954	1.8954
20	3.3539	3.3499

Table **1** depicts a comparative analysis of the present study with the existing literature. It has been observed that at $N_b = 0, N_t = 0, \lambda = 0, \gamma = \pi/2$ with the varying value of Pr, values of $\theta'(0)$ in the present problem are in decent promise with outcomes obtained by Khan and Pop [15].

In the current problem, we have considered the impact of various parameters like N_b, N_t, Pr, Le and variation of striking angle γ on the stretching sheet. Our goal is to study the heat and mass flux features of the problem. The effect of variations of the striking angle at specified parameters on dimensional velocity, dimensionless temperature, and concentration has been presented in Figs. (**2** to **4**) and Table **2**.

We have plotted the impact of variation in striking angle on velocity, temperature, and concentration profile keeping other parameters like M, λ, N_b, N_t, Pr and Le fixed in Figs. (**2**, **3**, and **4**), respectively. The respective values of constant parameters with the effect of different striking angles are shown in Table **2**. From Fig. (**2**), one can easily observe the effect of change in striking angle on velocity profile. It is observed that as the obliqueness increases, the local skin friction coefficients decrease. The respective values of local skin friction coefficients for various angles are shown in Table **2**. The cause for the decrease with obliqueness is that on lowering down of the angle, the wall shear stress decreases.

Table 2. Computed results for various fluid parameters.

Pr	γ	Le	Nb	Nt	$f''(0)$	$-\theta'(0)$	$-\phi'(0)$
10	$\pi/2$	10	0.1	0.1	−2.653930	1.149298	2.923452
10	$\pi/3$	10	0.1	0.1	−1.885775	1.116855	2.811603
10	$\pi/4$	10	0.1	0.1	−1.027422	1.077079	2.670854
10	$\pi/5$	10	0.1	0.1	−0.422444	1.046196	2.558014
10	$\pi/2$	10	0.1	0.05	−2.653930	1.381573	2.815040
10	$\pi/3$	10	0.1	0.05	−1.885775	1.341042	2.718118
10	$\pi/4$	10	0.1	0.05	−1.027422	1.291128	2.587129
10	$\pi/5$	10	0.1	0.05	−0.422444	1.252152	2.501068
10	$\pi/2$	10	0.1	0.01	−2.653930	1.606325	2.840395
10	$\pi/3$	10	0.1	0.01	−1.885775	1.557800	2.751813
10	$\pi/4$	10	0.1	0.01	−1.027422	1.497859	2.642091
10	$\pi/5$	10	0.1	0.01	−0.422444	1.450871	2.555791
10	$\pi/2$	10	0.05	0.1	−2.653930	1.539590	2.371427

(Table 2) cont.....

10	$\pi/3$	10	0.05	0.1	−1.885775	1.496006	2.250926
10	$\pi/4$	10	0.05	0.1	−1.027422	1.442531	2.096931
10	$\pi/5$	10	0.05	0.1	−0.422444	1.400973	1.971272
0.7	$\pi/2$	10	0.1	0.1	−2.653930	0.772897	2.762723
1	$\pi/2$	10	0.1	0.1	−2.653930	0.879042	2.738488
5	$\pi/2$	10	0.1	0.1	−2.653930	1.245104	2.729231
0.7	$\pi/3$	10	0.1	0.1	−1.885775	0.734266	2.676830
1	$\pi/3$	10	0.1	0.1	−1.885775	0.836591	2.652228
5	$\pi/3$	10	0.1	0.1	−1.885775	1.200532	2.631804
0.7	$\pi/5$	10	0.1	0.1	−0.422444	0.646544	2.486763
1	$\pi/5$	10	0.1	0.1	−0.422444	0.738736	2.460758
5	$\pi/5$	10	0.1	0.1	−0.422444	1.100373	2.412520
10	$\pi/2$	1	0.1	0.1	−2.653930	1.844488	1.888167
10	$\pi/2$	2	0.1	0.1	−2.653930	1.651119	0.624902
10	$\pi/2$	3	0.1	0.1	−2.653930	1.511430	1.132300
10	$\pi/2$	5	0.1	0.1	−2.653930	1.344601	1.834279
10	$\pi/2$	10	0.1	0.1	−2.653930	1.149298	2.923452
10	$\pi/5$	1	0.1	0.1	−0.422444	1.746430	0.298837
10	$\pi/5$	2	0.1	0.1	−0.422444	1.526067	0.395791
10	$\pi/5$	3	0.1	0.1	−0.422444	1.393347	0.864611

(Table 2) cont.....

10	$^{\pi}/_{5}$	5	0.1	0.1	−0.422444	1.233564	1.522039
10	$^{\pi}/_{5}$	10	0.1	0.1	−0.422444	1.046196	2.558014

From Fig. (**2**) and Table **2**, one can conclude that the Nusselt number changes with the change in striking angle. The Nusselt number rises with the obliqueness of the striking angle. It is due to the reason that the thermal boundary becomes thicker because of the angle of the incidence of the fluid. On the contrary, the Sherwood number rises with the obliqueness, which means the concentration boundary layer has a positive effect, thereby enhancing the concentration boundary layer. It can be seen in the concentration profile of (Fig. **4**) and Table **2**.

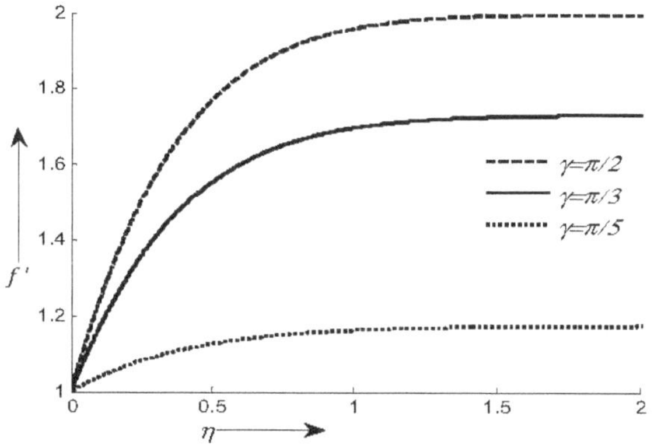

Fig. (2). Outcome of striking angle γ on velocity when $M = 3, Pr = 10 \lambda = 2$, $N_b = 0.1$, $N_t = 0.1$ and $Le = 10$.

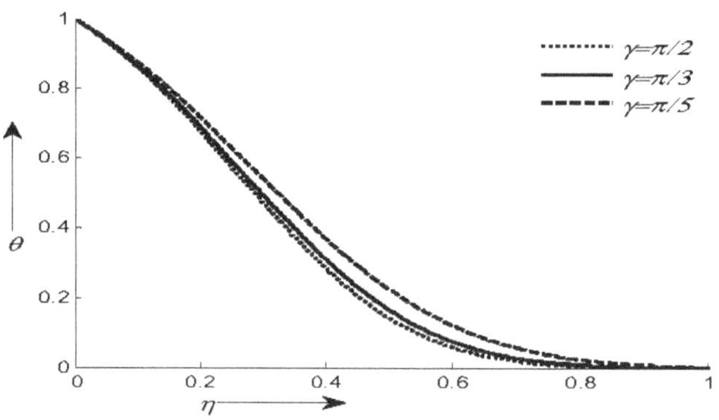

Fig. (3). Outcome of striking angle γ on temperature when $M = 3, Pr = 10 \lambda = 2$, $N_b = 0.1$, $N_t = 0.1$ and $Le = 10$.

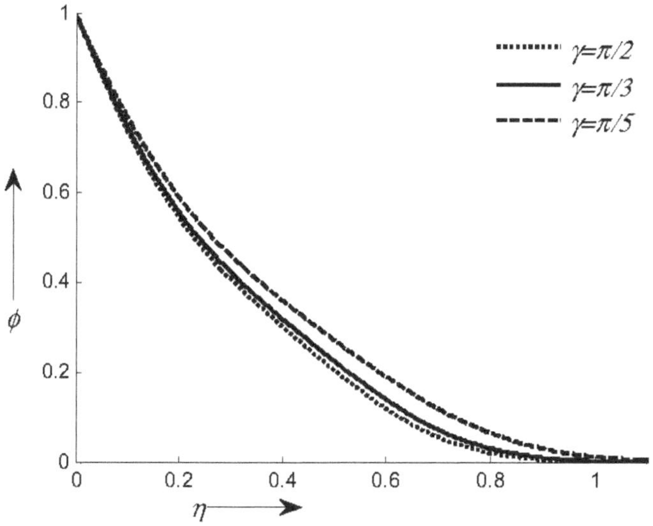

Fig. (4). Outcome of striking angle γ on concentration when $M = 3, Pr = 10 \lambda = 2$, $N_b = 0.1$, $N_t = 0.1$ and $Le = 10$.

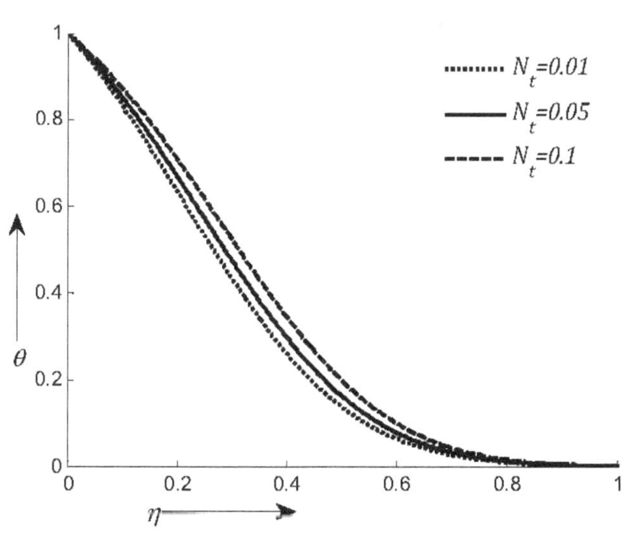

Fig. (5). Outcome of N_t on temperature when $M = 3, Pr = 10 \lambda = 2$, $N_b = 0.1$, $\gamma = \pi/4$ and $Le = 10$.

Figs. (**5** and **6**) depict the outcome of N_t at a certain fixed angle of incidence. The skin friction coefficient remains invariant with changes in the thermophoresis parameter but there is a variation in the local Nusselt number. As N_t increases from

0.01 to 0.1 then the Nusselt number increases whereas the reverse pattern is being followed by the Sherwood number. The behavior of dimensional temperature and the concentration distribution the for various thermophoresis parameters are plotted in Figs. (**5** and **6**).

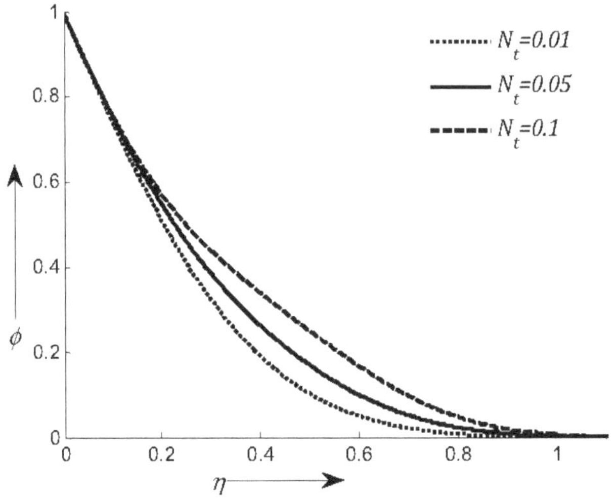

Fig. (6). Outcome of N_t on concentration when $M = 3, Pr = 10 \lambda = 2, N_b = 0.1, \gamma = \pi/4$ and $Le = 10$.

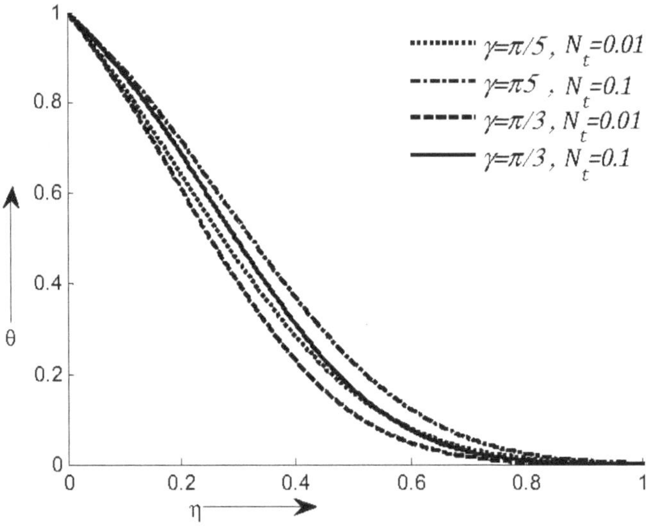

Fig. (7). Effect of thermophoresis parameter N_t and striking angle on temperature distribution when $M = 3, Pr = 10, \lambda = 2, N_b = 0.1$, and $Le = 10$.

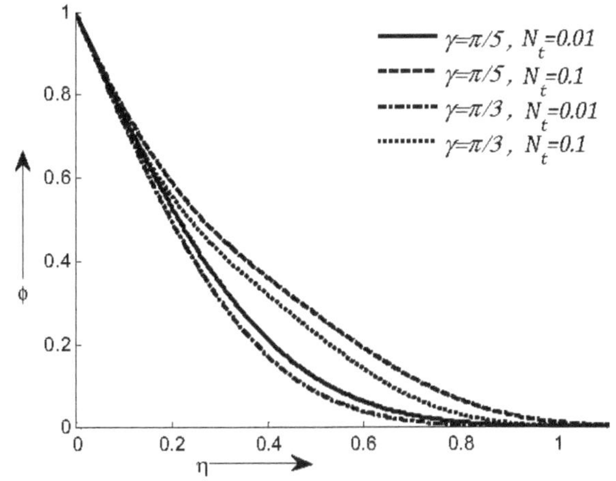

Fig. (8). Effect of thermophoresis parameter N_t and striking angle on concentration distribution when $M = 3, Pr = 10, \lambda = 2$, $N_b = 0.1$, and $Le = 10$.

Figs. (**7** and **8**) represent the combined influence of thermophoresis and the striking angle effect. They show that when we decrease the striking angle, there will be an associated growth in both dimensionless temperature and concentration. The outcome of the thermophoresis parameter as discussed above is that temperature and concentration decrease with a decrease in N_t. The combined effect of striking angle and thermophoresis gets enhanced as the striking angle decreases. So the reduced Nusselt number increases more with the rise in thermophoresis at the lower angle of incidence while the opposite of this happens with the local Sherwood number which gets more enhanced at lower angles.

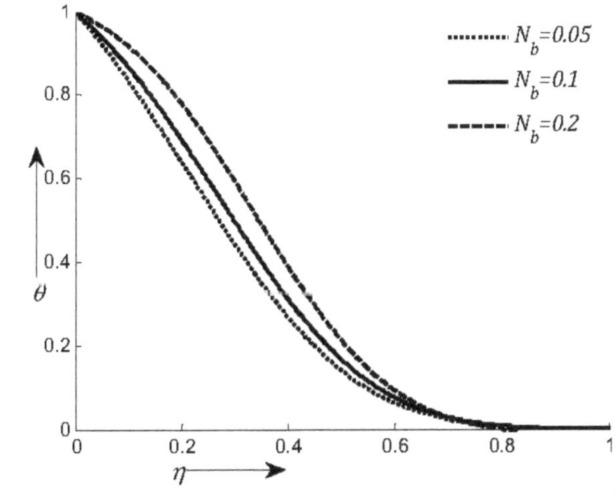

Fig. (9). Outcome of N_b on temperature distribution when $M = 3, Pr = 10, \lambda = 2$, $N_t = 0.1$, and $Le = 10$.

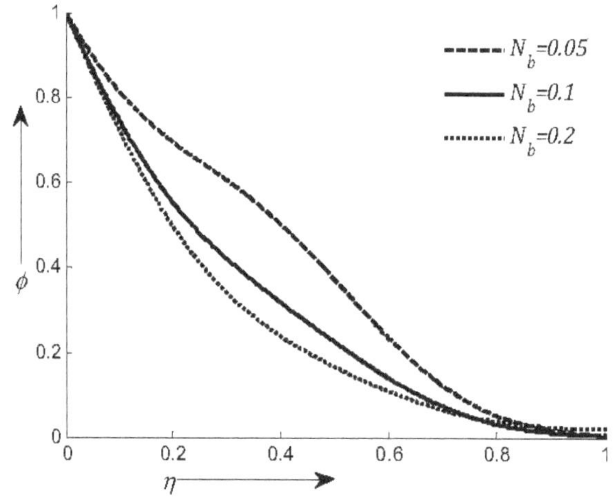

Fig. (10). Outcome of N_b on concentration when $M = 3, Pr = 10, \lambda = 2, N_t = 0.1$, and $Le = 10$.

We have listed the outcome of the change in N_b and its graphical representation has been presented in Figs. (**9** and **10**). It has been observed that as N_b rises then the temperature rises and the volume fraction profile declines. So the local Nusselt number rises with a rise in N_b whereas the local Sherwood number declines.

The graphical representations in Figs. (**11** and **12**) show that not only the variation in N_b but the change in striking angle affects the temperature and concentration profile. It has been noted that as the striking angle decreases then the temperature increases and the same result has been observed in the concentration profile also. So, both the Nusselt number and Sherwood number rise with a decrease in the striking angle. Figs. (**11** and **12**) show a comparative analysis of variation in both N_b and the striking angle. The reason for the change in concentration profile is the variation in the zigzag motion of the particles of the fluid. As the Brownian motion increases, there will be more random motion of fluid particles. Hence, they would strike more with each other, as a result, lots of heat would be produced due to these collisions. So Brownian motion heats the surface and hence concentration reduces due to the Brownian motion parameter.

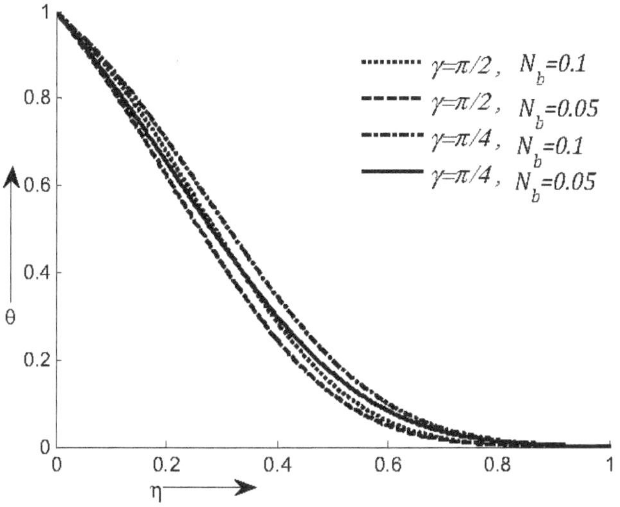

Fig. (11). Effect of N_b and striking angle on temperature distribution when $M = 3, Pr = 10, \lambda = 2$, $N_t = 0.1$, and $Le = 10$.

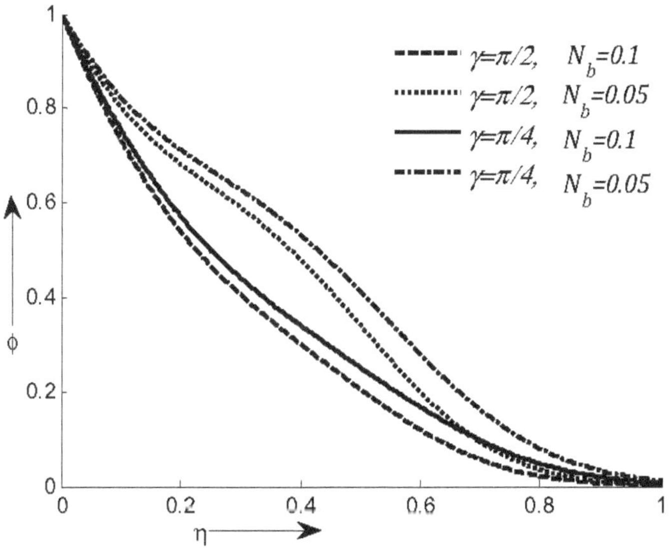

Fig. (12). Effect of N_b and striking angle on concentration distribution when $M = 3, Pr = 10, \lambda = 2, N_t = 0.1$, and $Le = 10$.

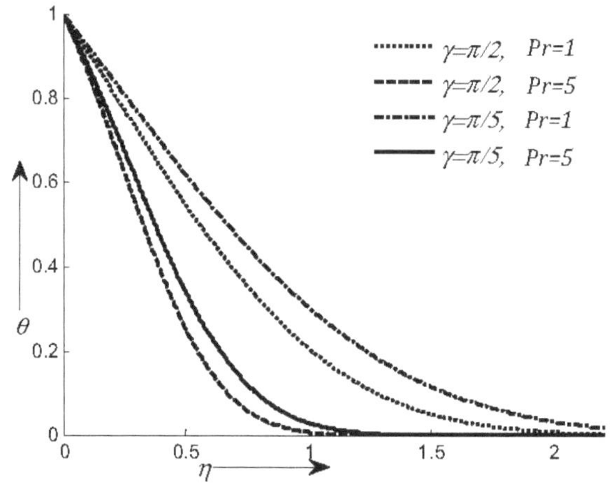

Fig. (13). Effect of Prandtl number Pr and striking angle on temperature distribution when $M = 3, N_b = 0.1, \lambda = 2, N_t = 0.1$, and $Le = 10$.

It is observed in Fig. (**13**) that as Pr increases, the dimensionless temperature decreases while in Fig. (**14**), the concentration distribution increases. The reason for this behavior is that as Pr increases, the thermal boundary layer becomes thinner and the temperature gradient rises.

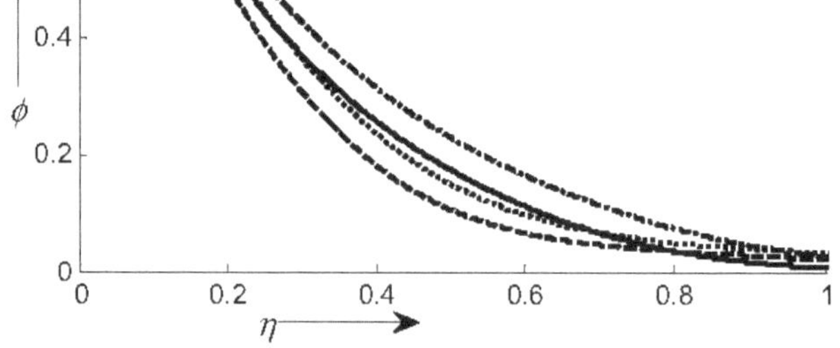

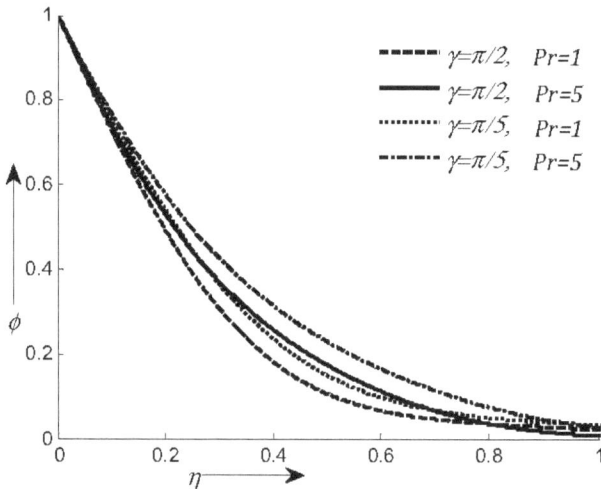

Fig. (14). Effect of Prandtl number Pr and striking angle on concentration distribution when $M = 3, N_b = 0.1, \lambda = 2, N_t = 0.1$, and $Le = 10$.

It has been observed that as the striking angle decreases, then the temperature distribution as well as the concentration distribution increases. Due to this, there will be an associated increase in the Nusselt number and a decrease in the Sherwood number.

The outcome of Le on temperature and concentration profile and the same has been presented graphically by Figs. (**15** and **16**). It has been noticed that as Le rises, the dimensionless temperature rises; on the other hand there is a decrease in the concentration distribution . The impact of decreasing the Lewis number is a decline in the temperature and hence; Nusselt number increases while the increasing concentration with a decreasing Lewis number leading to a decreasing Sherwood number.

It has been observed from graphical representations in Figs. (**15** and **16**) that change in the Lewis number and the striking angle affects the temperature and concentration profile. It can be noted as the striking angle reduces, the temperature increases, and the same effect is observed in the concentration profile.

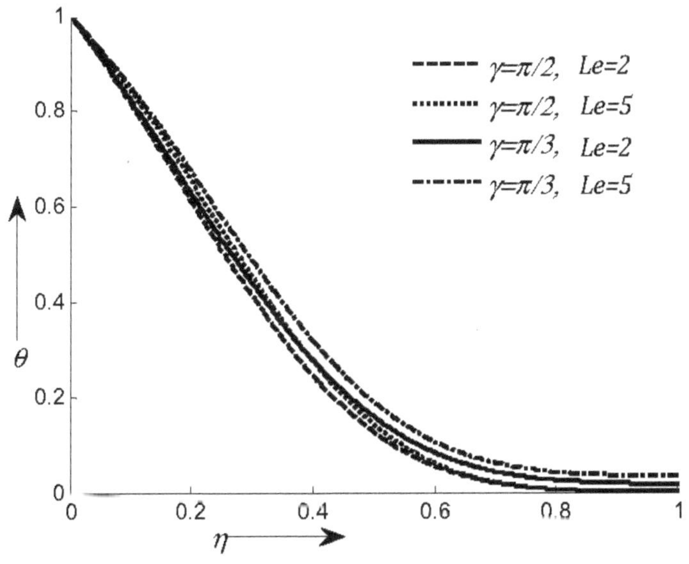

Fig. (15). Effect of Lewis numberLe and striking angle on temperature distribution when $M = 3, N_b = 0.1, \lambda = 2, N_t = 0.1$, and $Pr = 10$.

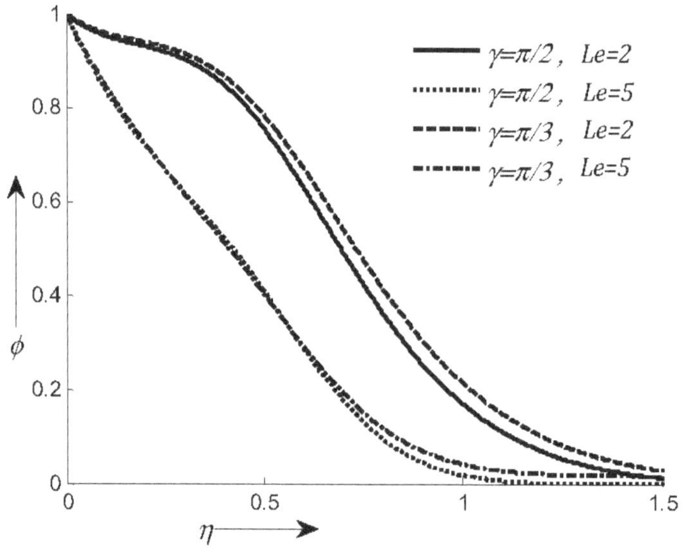

Fig. (16). Effect of Lewis numberLe and striking angle on concentration distribution when $M = 3$,$N_b = 0.1, \lambda = 2$, $N_t = 0.1$, and $Pr = 10$.

The streamlined outline for distinct striking angles is shown in Figs. (**17** to **20**) for $M = 0$.

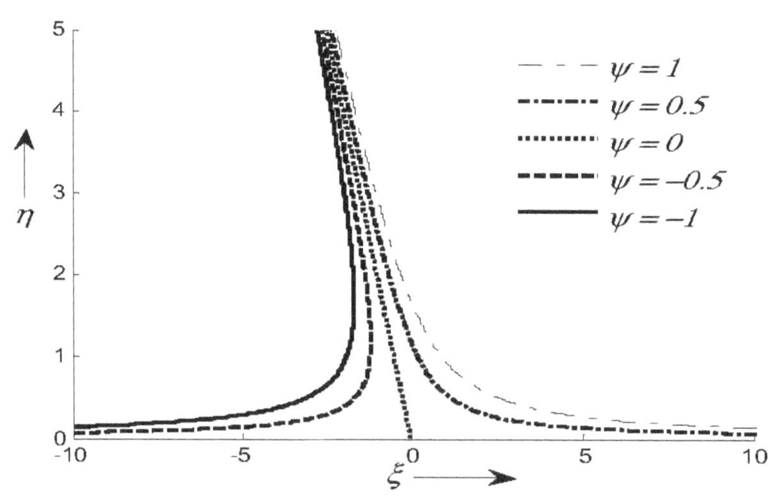

Fig. (17). Streamline outline when $\gamma = \pi/4$.

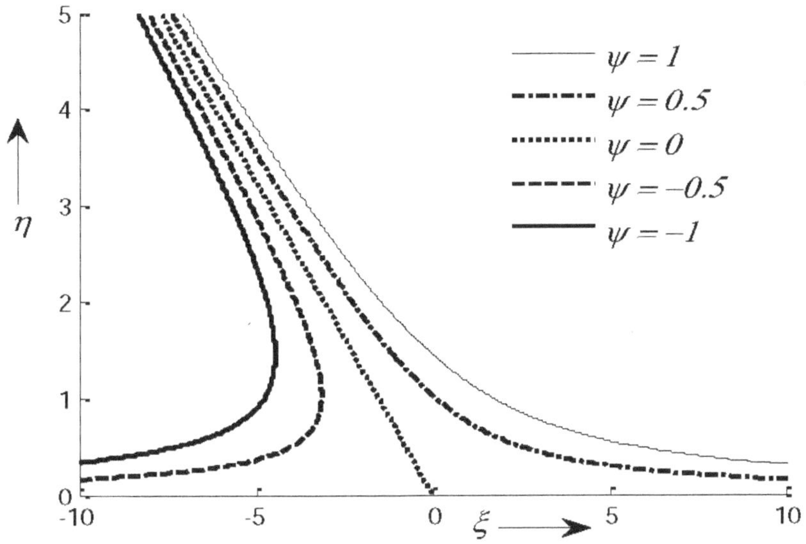

Fig. (18). Streamline outline when $\gamma = \pi/10$.

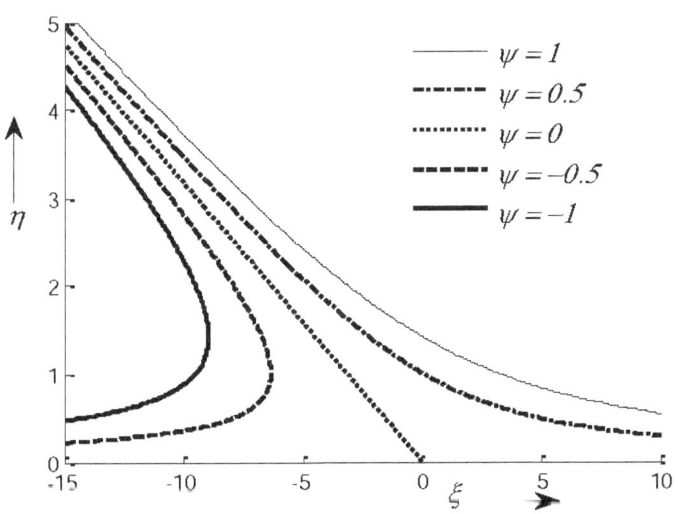

Fig. (19). Streamline outline when $\gamma = \pi/18$.

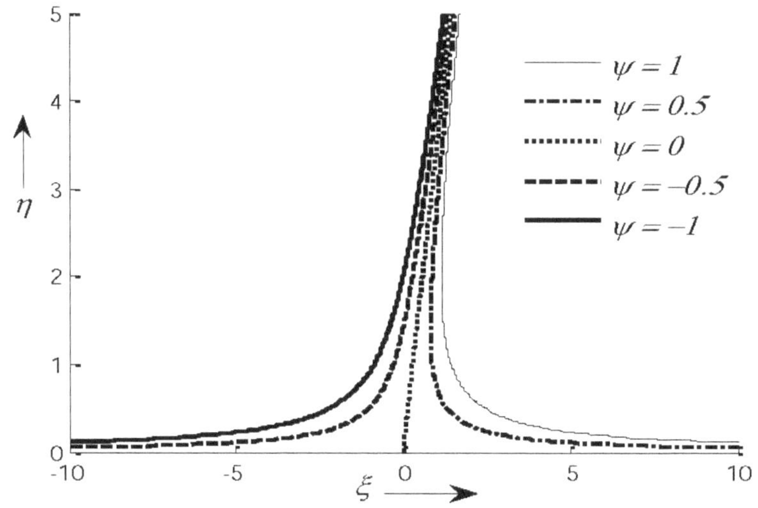

Fig. (20). Streamline outline when $\gamma = 2\pi/3$.

CONCLUSION

A constant 2-D flow of an incompressible viscous nanofluid has been examined when it strikes a stretched sheet at a certain angle of incidence. The outcomes of the distinct parameters with striking angle have been investigated. The following is a summary of the paper's main findings:

1. As the Brownian motion rises, the temperature rises.
2. As the Brownian motion rises, the nanoparticle concentration diminishes.
3. When the striking angle is reduced, the Brownian motion causes the temperature and concentration to rise.
4. With increasing N_t, the temperature and concentration decrease.
5. Temperature declines but the concentration rises as Le rises.
6. These results have important applications in stretching materials like liquid-based systems.

CONFLICT OF INTEREST

The author certifies that they have no financial or personal conflicts of interest or known personal ties that would have seemed to have an impact on the work disclosed in this publication.

REFERENCES

[1] S. U. S. Choi, and J. A. Eastman, "Enhancing thermal conductivity of fluids with nanoparticles", Argonne National Lab. (ANL), Argonne, IL (United States), ANL/MSD/CP-84938; CONF-951135-29, 1995.

[2] J. Buongiorno, "Convective transport in nanofluids", *J. Heat Transfer,* vol. 128, no. 3, pp. 240-250, 2006.
http://dx.doi.org/10.1115/1.2150834

[3] L.J. Crane, "Flow past a stretching plate", *Z. Angew. Math. Phys.,* vol. 21, no. 4, pp. 645-647, 1970.
http://dx.doi.org/10.1007/BF01587695

[4] T.R. Mahapatra, and A.S. Gupta, "Stagnation-point flow towards a stretching surface", *Can. J. Chem. Eng.,* vol. 81, no. 2, pp. 258-263, 2003.
http://dx.doi.org/10.1002/cjce.5450810210

[5] P. Singh, N.S. Tomer, and D. Sinha, "Oblique stagnation-point Darcy flow towards a stretching sheet", *J. Appl. Fluid Mech.,* vol. 5, pp. 29-37, 2012.

[6] P. Singh, and K. Tewari, "Non-Darcy free convection from vertical surfaces in thermally stratified porous media", *Int. J. Eng. Sci.,* vol. 31, no. 9, pp. 1233-1242, 1993.
http://dx.doi.org/10.1016/0020-7225(93)90128-H

[7] P. Singh, N.S. Tomer, S. Kumar, and D. Sinha, "MHD oblique stagnation-point flow towards a stretching sheet with heat transfer", *Int. J. Appl. Math. Mech.,* vol. 6, pp. 94-111, 2010.

[8] P. Singh, T. Singh, S. Kumar, and D. Sinha, "Effect of radiation and porosity parameter on magnetohydrodynamic flow due to stretching sheet in porous media", *Therm. Sci.,* vol. 15, no. 2, pp. 517-526, 2011.
http://dx.doi.org/10.2298/TSCI1102517S

[9] M.A.A. Hamad, "Analytical solution of natural convection flow of a nanofluid over a linearly stretching sheet in the presence of magnetic field", *Int. Commun. Heat Mass Transf.,* vol. 38, no. 4, pp. 487-492, 2011.
http://dx.doi.org/10.1016/j.icheatmasstransfer.2010.12.042

[10] N. Bachok, A. Ishak, and I. Pop, "Stagnation-point flow over a stretching/shrinking sheet in a nanofluid", *Nanoscale Res. Lett.,* vol. 6, no. 1, pp. 623-628, 2011.
http://dx.doi.org/10.1186/1556-276X-6-623 PMID: 22151965

[11] N. Bachok, A. Ishak, and I. Pop, "Boundary layer stagnation-point flow toward a stretching/shrinking sheet in a nanofluid", *J. Heat Transfer,* vol. 135, no. 5, p. 054501, 2013.
http://dx.doi.org/10.1115/1.4023303

[12] J.H. Holland, Adaptation in Natural and Artificial Systems, University of Michigan Press, Ann Arbor, Michigan, 1975; re-issued by MIT Press 1992.

[13] J. McCall, "Genetic algorithms for modelling and optimisation", *J. Comput. Appl. Math.,* vol. 184, no. 1, pp. 205-222, 2005.
http://dx.doi.org/10.1016/j.cam.2004.07.034

[14] N. Chaiyaratana, and A.M.S. Zalzala, "Hybridisation of neural networks and genetic algorithms for time-optimal control", In Proceedings of the 1999 Congress on Evolutionary Computation-CEC99 (Cat. No. 99TH8406), Washington, DC, USA, 6–9 July 1999.
http://dx.doi.org/10.1109/CEC.1999.781951

[15] W.A. Khan, and I. Pop, "Boundary-layer flow of a nanofluid past a stretching sheet", *Int. J. Heat Mass Transf.,* vol. 53, no. 11-12, pp. 2477-2483, 2010.
http://dx.doi.org/10.1016/j.ijheatmasstransfer.2010.01.032

CHAPTER 5

Effect of Casson Nanofluid's Outer Velocity with Melting Heat Transmission beyond a Stretching Sheet

Vikas Poply[1,*], Sabyasachi Mondal[2] and Parveen Kumar[3]

[1] *Department of Mathematics, KLP College Rewari, Haryana, India*

[2] *Department of Mathematics, North- Eastern Hill University, (NEHU) Shillong, Meghalaya, India*

[3] *Department of Physics, Govt. P.G. Nehru College, Jhajjar, Haryana, India*

Abstract: An analysis was conducted to observe the collective consequence of the melting procedure across a stretching sheet in a Casson nanofluid with outer velocity. The outcomes of melting heat parameter (M) and Casson parameter (β) are shown in this analysis. The leading PDEs are moulded in ODE with similarity variables. The numerical results of moulded ODE are given by the RKF scheme with shooting procedures. Important parameters of the nanofluid are examined and graphically displayed. Tables are used to display the Sherwood number, Nusselt number, and skin friction coefficient based on different nanofluid characteristics. The development of cooling systems that raise the heat transmission belongings of nanofluids and increase engineering potential will be greatly aided by this research.

Keywords: Casson nanofluid, Melting heat transmission, Outer velocity.

INTRODUCTION

The classic boundary layer issues are expressed as PDEs. It is commonly recognized that PDEs are used in science and technology fields. The studies [1, 2] broadly explain their foremost belongings on real-world challenges.

Based on the investigational data, nanofluids are more thermally conductive than ordinary fluids. Eastman and Choi developed nanofluids back in 1995 who argued that the existence of scattered nanoparticles increased the pure fluid's true thermal conductivity [3]. In nanofluids, the depth coverage of convective transference was provided by Buongiorno [4].

*Corresponding author Vikas Poply:** Department of Mathematics, KLP College Rewari, Haryana, India; E-mail: vikaspoply@gmail.com

It is common knowledge that physical transformation takes on melting by heating which results in the phase transition of materials from solid to liquid. The melting process has served as the subject of incredibly challenging theoretical and practical research for many years. Real-world applications that include welding and casting in manufacturing, defrosting frozen ground, conserving heat energy, heating, and cooling.

Consequently, melting has been used in a few experiments. The melting of ice in an airstream that is heated was studied by Roberts [5]. The relationship between melting and diffusion transmission of heat was primarily studied [6]. In laminar flow, the transmission of melting heat was studied by Epstein and Cho [7]. The use of natural convection for melting was studied by Sparrow *et al.* [8]. Only a few researchers [9-13], have recently researched melting heat transmission in boundary layer problems with various angles.

Therefore, heat transmission *via* melting in a Casson nanofluid flow past a stretched sheet with an outside flow was investigated in this work. The method of Newton-Fehlberg iteration is used to crack the transformed nondimensional ODE. Detailed descriptions of the pertinent flow factors are provided.

MATHEMATICAL FORMULATION

Consider a stretched sheet that is implanted through outside velocity and melts while being continuously traversed through a 2-D incompressible Casson nanofluid, as depicted in Fig. (**1**).

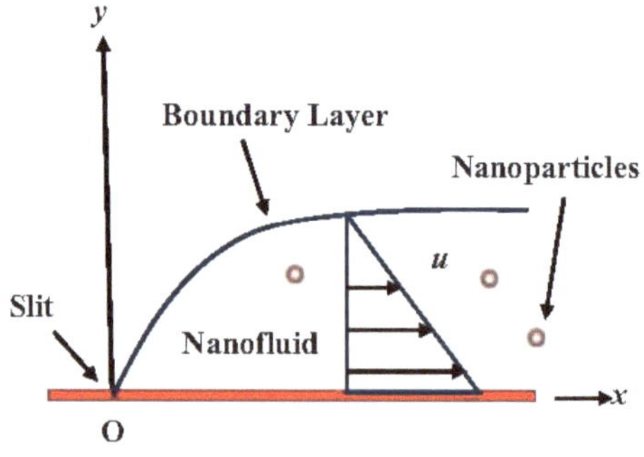

Fig. (1). Schematic diagram.

The governing equations are [14]:

$$\frac{\partial u}{\partial x} + \frac{\partial v}{\partial y} = 0 \quad \textbf{(1)}$$

$$u\frac{\partial u}{\partial x}+v\frac{\partial u}{\partial y}=U(x)\frac{\partial U(x)}{\partial x}+v\left(1+\frac{1}{\beta}\right)\frac{\partial^2 u}{\partial y^2} \quad (2)$$

$$u\frac{\partial T}{\partial x}+v\frac{\partial T}{\partial y}=\alpha\frac{\partial^2 T}{\partial x^2}+\tau\left\{D_B\frac{\partial C}{\partial y}\frac{\partial T}{\partial y}+\frac{D_T}{T_\infty}\left(\frac{\partial T}{\partial y}\right)^2\right\} \quad (3)$$

$$u\frac{\partial C}{\partial x}+v\frac{\partial C}{\partial y}=D_B\frac{\partial^2 C}{\partial y^2}+\frac{D_T}{T_\infty}\frac{\partial^2 T}{\partial y^2} \quad (4)$$

Restricted surroundings designed for the flow are:

$$v=0,\quad u=u_w,\ k\frac{\partial T}{\partial y}=\rho[LH+C_s(T_w-T_0)]v(x,0),\ T=T_w,\ C=C_w \quad at\ y=0 \quad (5)$$

$$u\to U(x),\qquad v=0,\qquad T\to T_\infty\qquad C\to C_\infty\qquad as\ y\to\infty \quad (6)$$

In this case, thermal conductivity, solid surface temperature, latent heat, heat capacity and density are represented by variables K, T_0, LH, C_s and ρ

The dimensionless and similarity quantities are:

$$\psi=\sqrt{xvU(x)}\,f(n),\quad \eta=\sqrt{\frac{U(x)}{xv}}y,\quad \theta(\eta)=\frac{T-T_W}{T_\infty-T_W},\quad \phi(\eta)=\frac{C-C_\infty}{C_W-T_\infty} \quad (7)$$

Stream function $\psi(x,y)$ is given as:

$$v=-\frac{\partial\psi}{\partial x},\ u=-\frac{\partial\psi}{\partial x} \quad (8)$$

Using Eqs. (7) and (8), Eqs. (2-4) become:

$$\left(1+\frac{1}{\beta}\right)f'''+(\lambda^2-f'^2)+ff''=0 \quad (9)$$

$$\theta''+P_r\{N_b\phi'\theta'+f\theta'+N_t\theta'^2\}=0 \quad (10)$$

$$\phi''+\frac{N_t}{N_b}\theta''+Le\phi'f=0 \quad (11)$$

The restricted surroundings are as follows:

$$B\theta'(\eta)+P_rf(\eta)=0,\ f'(\eta)=0\ \ \theta(\eta)=0,\quad \phi(\eta)=0 \text{ at } \eta=0 \quad (12)$$

$$f'(\eta) \to \lambda \quad \theta(\eta) \to 1, \;\; \phi(\eta) \to 1 \;\; \text{at}\, \eta = \infty, \tag{13}$$

The physical quantities of our interest are as:

$$\sqrt{Re_x}Cf_x = \left(1 + \frac{1}{\beta}\right) f''(0), \quad \frac{Nu}{\sqrt{Re_x}} = -\theta'(0), \;\; \frac{Sh}{\sqrt{Re_x}} = -\phi'(0) \tag{14}$$

DISCUSSION AND OUTCOMES

Using the Newton-Fehlberg iteration approach, the transformed nondimensional equations with boundary restrictions are computed. The belongings of melting heat parameter (M) and Casson parameter (β) are studied in detail. Table **1** represents the comparison of $-\theta'(0)$ for different values of β and validation of results is done by Khan and Pop [15].

The outcomes of β on $f'(\eta), \theta(\eta)$ and $\phi(\eta)$ are depicted in Figs. (**2**, **3**, and **4**). It is evident from figures that as β increases, temperature and concentration increase while velocity decrease. The calculated outcomes for $f''(0), -\theta'(0)$ and $-\phi'(0)$ for distinct entries of β are shown in Table **2**.

Table 1. $-\theta'(0)$ **for different** Pr**, a comparison.**

Pr	**Khan and Pop [15]**	**Present results**
0.7	0.4539	0.4544
2	0.9114	0.9113
7	1.8954	1.8954

Table 2. Outcomes of $f''(0)$, $-\theta'(0)$ **and** $-\phi'(0)$ **for** β**.**

β	$f''(0)$	$-\theta'(0)$	$-\phi'(0)$
1	−0.7072	0.2728	0.5404
2	−0.8165	0.2646	0.5059
3	−0.8661	0.2608	0.4909

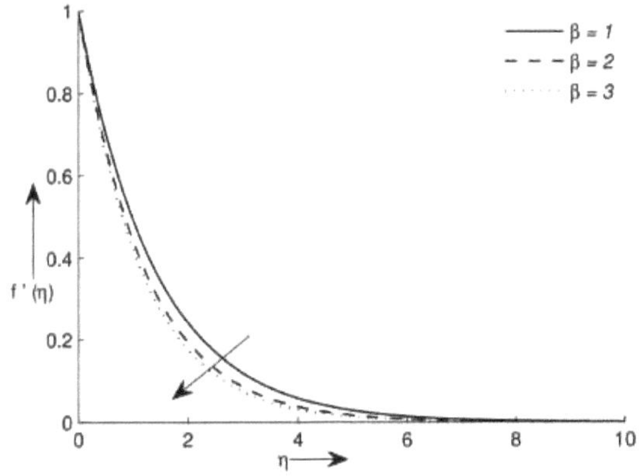

Fig. (2). $f'(\eta)$ for several β.

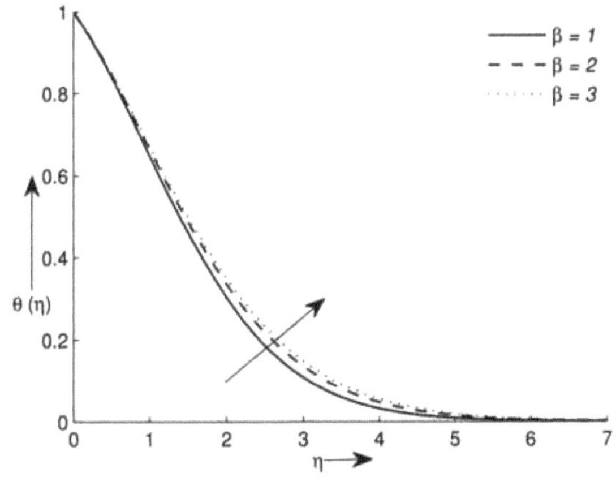

Fig. (3). $\theta(\eta)$ for several β.

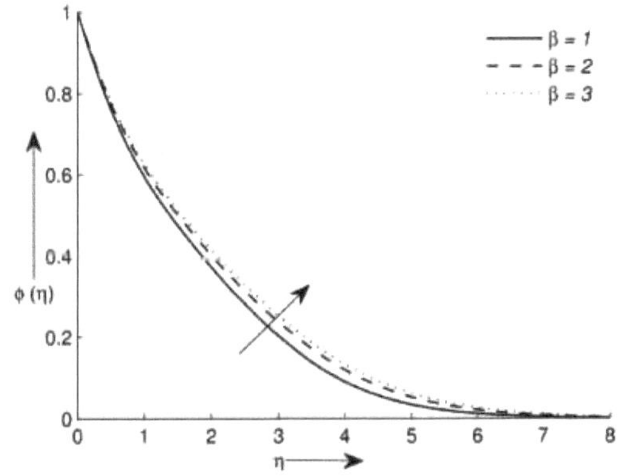

Fig. (4). $\phi(\eta)$ for several β.

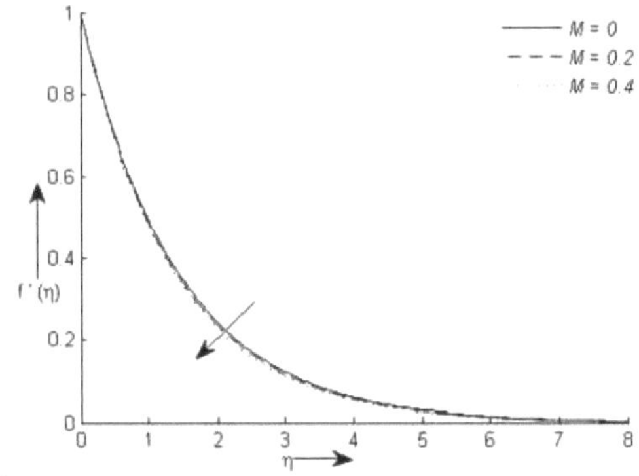

Fig. (5). $f'(\eta)$ for several M.

The outcomes of M on $f'(\eta), \theta(\eta)$ and $\phi(\eta)$ are depicted in Figs. (**5**, **6**, and **7**). From Figs., all things decline with a rise in M because of higher M leads in more molecular mobility, which in turn causes energy to be dissipated and the fluid's temperature to drop. Table **3** describes the calculated outcomes for $f''(0), -\theta'(0)$ *and* $-\phi'(0)$ for diverse M.

Table 3. Calculated outcomes of $f''(0)$, $-\theta'(0)$ and $-\phi'(0)$ for M.

M	$f''(0)$	$-\theta'(0)$	$-\phi'(0)$
0	−0.7072	0.2728	0.5404
0.2	−0.7218	0.2893	0.5620
0.4	−0.7387	0.3086	0.5867

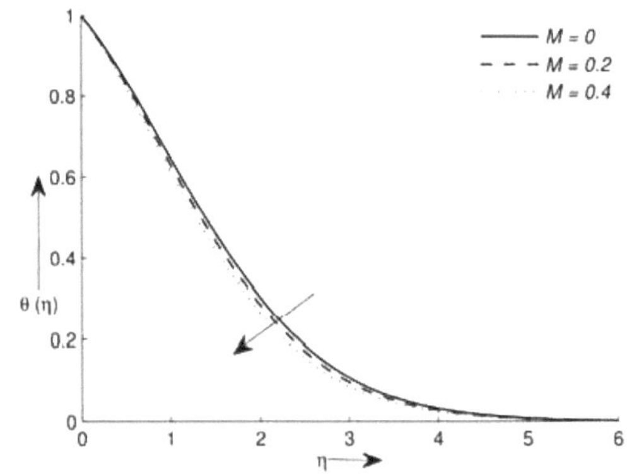

Fig. (6). $\theta(\eta)$ for several M.

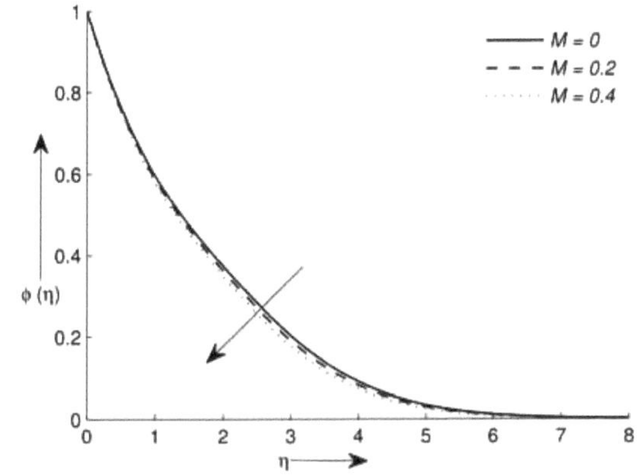

Fig. (7). $\phi(\eta)$for several M.

CONCLUSION

From the outcomes the results we can conclude the following:

1. As β increases, concentration and temperature rise and velocity reduces.
2. Temperature, concentration, and movement all reduces as M rises.

REFERENCES

[1] M. Ghergu, and V. Radulescu, *Nonlinear PDEs:* "Mathematical Models in Biology, Chemistry and Population Genetics", *Springer Monographs in Mathematics.,* Springer: Heidelberg, 2012. http://dx.doi.org/10.1007/978-3-642-22664-9

[2] M.M. Cavalcanti, V.N. Domingos Cavalcanti, I. Lasiecka, and C.M. Webler, "Intrinsic decay rates for the energy of a nonlinear viscoelastic equation modeling the vibrations of thin rods with variable density", *Adv. Nonlin. Analy.,* vol. 6, no. 2, pp. 121-145, 2017. http://dx.doi.org/10.1515/anona-2016-0027

[3] S. Chol, and J. Estman, "Enhancing thermal conductivity of fluids with nanoparticles", *ASME-Publications-Fed.,* vol. 231, pp. 99-106, 1995.

[4] J. Buongiorno, "Convective transport in nanofluids", *J. Heat Transfer,* vol. 128, no. 3, pp. 240-250, 2006. http://dx.doi.org/10.1115/1.2150834

[5] L. Roberts, "On the melting of a semi-infinite body of ice placed in a hot stream of air", *J. Fluid Mech.,* vol. 4, no. 5, pp. 505-528, 1958. http://dx.doi.org/10.1017/S002211205800063X

[6] Y.C. Yen, and C. Tien, "Laminar heat transfer over a melting plate, The modified leveque problem", *J. Geophys. Res.,* vol. 68, no. 12, pp. 3673-3678, 1963. http://dx.doi.org/10.1029/JZ068i012p03673

[7] M. Epstein, and D.H. Cho, "Melting heat transfer in steady laminar flow over a flat plate", *J. Heat Transfer,* vol. 98, no. 3, pp. 531-533, 1976.
http://dx.doi.org/10.1115/1.3450595

[8] E.M. Sparrow, S.V. Patankar, and S. Ramadhyani, "Analysis of melting in the presence of natural convection in the melt region", *J. Heat Transfer,* vol. 99, no. 4, pp. 520-526, 1977.
http://dx.doi.org/10.1115/1.3450736

[9] A. Ishak, R. Nazar, N. Bachok, and I. Pop, "Melting heat transfer in steady laminar flow over a moving surface", *Heat Mass Transf.,* vol. 46, no. 4, pp. 463-468, 2010.
http://dx.doi.org/10.1007/s00231-010-0592-8

[10] W. Ibrahim, "Magnetohydrodynamic (MHD) boundary layer stagnation point flow and heat transfer of a nanofluid past a stretching sheet with melting", *Propulsion and Power Research,* vol. 6, no. 3, pp. 214-222, 2017.
http://dx.doi.org/10.1016/j.jppr.2017.07.002

[11] R.G. Abdel-Rahman, M.M. Khader, and A.M. Megahed, "Melting phenomenon in magneto hydro-dynamics steady flow and heat transfer over a moving surface in the presence of thermal radiation", *Chin. Phys. B,* vol. 22, no. 3, p. 030202, 2013.
http://dx.doi.org/10.1088/1674-1056/22/3/030202

[12] S. Ahmad, and I. Pop, "Melting effect on mixed convection boundary layer flow about a vertical surface embedded in a porous medium: opposing flows case", *Transp. Porous Media,* vol. 102, no. 3, pp. 317-323, 2014.
http://dx.doi.org/10.1007/s11242-014-0291-x

[13] N.A. Yacob, A. Ishak, and I. Pop, "Melting heat transfer in boundary layer stagnation-point flow towards a stretching/shrinking sheet in a micropolar fluid", *Comput. Fluids,* vol. 47, no. 1, pp. 16-21, 2011.
http://dx.doi.org/10.1016/j.compfluid.2011.01.040

[14] V. Makkar, and V. Poply, "Impact of outer velocity on flow, heat and mass transfer of Casson nanofluid over a non-linear stretching sheet", *J. Therm. Eng.,* vol. 7, pp. 1353-1365, 2021.

[15] W.A. Khan, and I. Pop, "Boundary layer flow of a nanofluid past a stretching sheet", *Int. J. Heat Mass Transf.,* vol. 53, no. 11-12, pp. 2477-2483, 2010.
http://dx.doi.org/10.1016/j.ijheatmasstransfer.2010.01.032

CHAPTER 6

Dual Solutions of Magneto-Radiative Ag-Water Nanofluid Slip Flow Across the Porous Medium Due to a Permeable Contracting Surface with Heat Generation: Stability Analysis

Gopinath Mandal[1,*]

[1] *Siksha-Satra, Sriniketan, Visva-Bharati University, West Bengal, India*

Abstract: The current study looks at the heat transfer of a magneto-radiative nanofluid over an exponentially contracting permeable sheet in the existence of heat generation with numerous slip boundary constraints. It also looks at the stability and duality of solutions. Here, a water-based fluid with silver (Ag) nanoparticles is employed. Before being computed by bvp4c in the Matlab program, the governing nonlinear partial differential equations are converted into dimensionless nonlinear ODEs *via* a similarity transformation. Due to the contracting surface scenario, a dual-nature solution can only be found if a sufficient suction value is used. We may infer from stability investigation that the first one is stable while the second one is unstable. A stable solution makes sense with the least positive eigenvalues, but a lower unstable solution indicates negative eigenvalues. The Nusselt number and the skin-friction factor may be improved by increases in the silver nanoparticle's solid volume percentage. The least eigenvalue converges to zero as the suction and contracting surface parameters reach their critical values. Nanofluids that are magnetic and radiative can be used to create beautiful and effective electromagnetic devices, including in the areas of clothing, paper, plastics, food colorants, cars, cancer therapy, medicines, ceramics, soaps, and paints.

Keywords: Dual solutions, Exponentially contracting surface, Heat generation, Magnetic field, Nanofluid, Porous medium, Stability analysis, Thermal emission.

INTRODUCTION

Energy conservation in heat transmit systems is a critical issue in a variety of cutting-edge industries and mechanical applications. Conservative fluids such as water, polymeric solutions, biofluids, glycols, and lubricants, have been employed as heat transmit fluids for many years. Due to their poor thermal conductivity, they have a constrained ability to transport heat. The conductivity of nanofluids, which

*Corresponding author Gopinath Mandal: Siksha-Satra, Sriniketan, Visva-Bharati University, West Bengal, India; E-mail: gopi_math1985@rediffmail.com

is a combination of nano-sized particles-which may be metal oxides, metals, polymers, or other materials of size 1-100 nm-and the host fluid is greater, increasing the rate of heat transmission. Choi and Eastman [1] first conducted several studies to validate their novel, ground-breaking concept. Applying metal flakes with high thermal conductivities requires only a small amount of the particles. Studies on heat transmission in nanofluid movement over contracting surfaces have gained popularity due to their practical applications in a variety of technical and industrial disciplines, including the manufacture of paper, artificial fiber and polymer extraction, extrusion, and glass blowing. Researchers have also begun to apply stability analyses to issues with dual contracting-surface solutions, with one being steady and the other unstable. The first stability study to verify the stable solution was performed by Merkin [2]. When calculating eigenvalues, he discovered that nonnegative eigenvalues had stable while the negative eigenvalues had an inconsistent solution. Later, a large number of scholars performed stability analyses for dual solutions in mixed convection movements or extending or contracting surfaces. Weidman *et al.* [3] investigated the impacts of transpiration on movement on moving sheets. They discovered multiple findings and stated that, when blowing and suction effects are taken into account, the spectrum of dual solutions either diminishes or grows. Dero *et al.* [4] investigated slip impacts on the unsteady transport of nanofluid across a contracting surface while using several findings. Ghosh and Mukhopadhyay [5] experimented with nanofluid movement through an exponentially penetrable shrinking sheet in the existence of slip, with a dual nature of solution and stability analysis. The creation of entropy by nonlinear heat emission on a non-Newtonian nanofluid in magnetohydrodynamics was studied numerically by Bhatti *et al.* [6] using a permeable contracting sheet. Tiwari *et al.* [7] studied MHD flow across a non-linearly extending or contracting sheet using water-based nanofluids made of silver (Ag) and titanium (TiO_2). Triple solutions were discovered by Lanjwani *et al.* [8], who looked into the heat transmission properties of mixed-convective Ag-H_2O nanofluid flow across a permeable contracting and extending sheet. Fe_2O_3 and Fe-water base nanofluid boundary layer movement and heat transmission across an extending or contracting sheet with a radiation impact were studied by Lanjwani *et al.* [9] in terms of stability. The stability study for a hybrid nanofluid movement for the effects of Joule heating and viscosity across a contracting plate was conducted by Rasool *et al.* [10]. The stability study of radiative magnetic hybrid nanofluid slip movement caused by an exponentially extending or contracting penetrable sheet with heat production was recently the topic of research by Mandal and Pal [11].

The mixing of electromagnetism with fluid mechanics, or the effects of magnetic fields on electrically conducting fluids that can regulate the rate of heat transmission and flow in a system, is studied as magnetohydrodynamics. A magnetic field causes a fluid to travel across it, creating an electric current. In a variety of engineering and industrial applications, such as in nuclear reactors, magnetic mixers, MHD power generators, chemical reactions, metal casting, petroleum industries, and metallurgical processes, this electric current bears the potential to significantly affect both the flow and temperature behavior of the fluid. Using suction and injection, heat generation and absorption, and a heat-permeable extending and contracting surface, Chaudhary and Kanika [12] studied the influence of hydromagnetic forces on the movement of radiative nanofluids having stagnation points. The unsteady MHD stagnation point movement for nanofluid *via* the penetrable contracting surface has been examined by Roy and Pop [13]. Casson nanofluid across a vertical exponentially diminishing sheet having an angled magnetic field was examined by Ishtiaq *et al.* [14]. A penetrable stretched sheet with heat emission and viscous dissipation was the object of investigation in magnetohydrodynamic nanofluid movement by Muntazir *et al.* [15]. Using slip boundary constraints and entropy generation, Mandal and Pal [16] studied the radiative magneto single and multi-walled CNT/H_2O hybrid nanofluid flow.

Many recent studies have been published and are fully available on the subject of heat emission over the motion of the boundary layer. Convective flows in nanofluids are also extensively used in engineering, such as in geothermal systems, groundwater pollution, nuclear waste storage, crude oil extraction, thermal insulation, agriculture, heat exchangers, refrigerators, hybrid engines, microelectronics, cancer therapy, vehicle air conditioning systems in transportation, *etc*. The flowing movement of micropolar nanofluid *via* the stretched surface having a heat sink or source (non-uniform) was shown by Pal and Mandal [17] using thermal emission and MHD processes. The effects of heat emission on nanofluid movement bound by partial slip constraints were investigated by Rusdi *et al.* [18]. Azam [19] examined the importance of heat flow and non-linear thermal emission over an Arrhenius-activated magnetohydrodynamic Maxwell nanofluid. An issue with the calculation of thermal emission in nonlinear mode over magneto nanofluid movement by entropy production was shown by Farooq *et al.* in their study [20]. Alzahrani [21] investigated the impact of thermal emission on heat transmission in a Casson nano liquid containing suction flowing in a plane wall jet under slip boundary constraints.

Motivated by the preceding work, we can state that very few experiments were performed to calculate heat transmission and stability analysis in the existence of a magnetic field with thermal emission for *Ag*-water nanofluid influenced by permeable medium over the exponentially contracting surface. They did not consider the velocity and thermal slip in nanofluids and their impact on the boundary surface with mass suction. The main goal of the present work is to examine the magnetic-radiative silver-water nanofluid movement across a permeable medium driven by an exponentially contracting surface having thermal and velocity slip in the presence of sufficient mass suction and a heat source to study the separation in the boundary layer flow. Water is employed as the base liquid, and silver (Ag) is chosen as the nanoparticle due to its widespread use in research. Good chemical stability, catalytic and antibacterial abilities, as well as cytotoxic effects on cancer cells, are all shown to be characteristics of silver nanoparticles. They also display good electrical conductivity. The rate of heat transmission, temperature, wall shear stress, and velocity are all carefully investigated. The outcomes are tallied, and the identical data is shown visually. The bvp4c tool in MATLAB is used to do mathematical calculations to solve the nonlinear flow problem and carry out the stability investigation. The results of this study are innovative and haven't been observed by other researchers, as far as we know. Due to very complex partial differential and ordinary equations, this is theoretically feasible. We believe that we have thoroughly researched this issue. The following innovative aspects of the current study are highlighted:

- Introduction of *Ag*-water nanofluid movement through permeable medium subject to the exponentially shrieked sheet.
- Consideration of external magnetic, thermal emission, and heat generation in thermal management.
- Consideration of second-order velocity and thermal slip boundary constraints under the consideration of the nanofluid.
- Consideration of boundary layer separation points for sufficient muss suction and proper contracting velocity to control the flow efficiently.
- Using the MATLAB bvp4c approach to find dual solutions for the problem's highly nonlinear equations.

It is therefore expected that the answers to this flow problem can be used in a variety of engineering-industrial purposes, for example, blood flows, glass blowing, lubrication, electronic chips, foodstuffs, slurries, *etc*. This research is the first of its kind and contains various instances of how nanofluids are now utilized in industry.

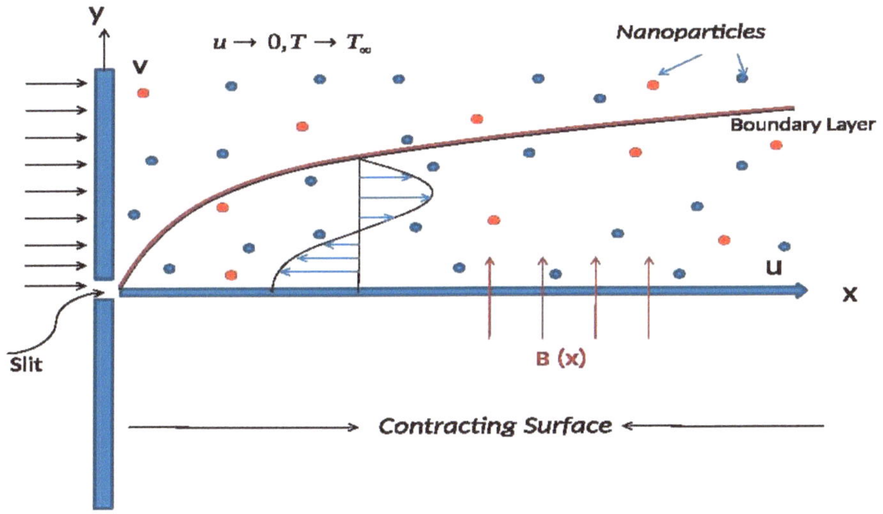

Fig. (1). Schematic flow diagram.

MATHEMATICAL FORMULATION

The investigation of a steady, incompressible boundary layer slips the movement of Ag water nanofluid towards an exponentially permeable contracting surface with thermal conductivity and here heat production is scrutinized. It is considered that the nanofluid flows in the x-direction with velocity u and in the y-direction with velocity v (Fig. **1**). The velocity of the shrinkable sheet is $u_w(x) = ce^{x/L}$, whereas constant mass velocity $v_w(x) = v_0 e^{x/2L}$. Also, it is presumed that variable wall temperature $T_w = T_\infty + T_0 e^{x/2L}$ $(T_0 > 0, T_w > T_\infty)$ presents a heated sheet and *vice versa*. Extension, contraction, and the motionless surface are represented by $\lambda >, <, = 0$. In addition, an applied magnetic field $\hat{B} = B_0 e^{x/2L}$ with content magnetic field strength B_0 is applied normally to the movement surface $y = 0$. Because the magnetic force by induction and impressed electric field are ignored. The thermal characteristics of nanofluids fluctuate dramatically when temperature, nanoparticle type, pressure, and other factors change. Lastly, the nanofluid is assumed to be optically thick, radiative heat transmission is addressed, and the Rosseland approximation is used to account for radiation effects. It is believed that nano-like particles are considered to be in thermal equilibrium with the base liquid. Considering these assumptions, the boundary layer equations including the boundary constraints (Wahid *et al.* [22]) are written as:

$$\frac{\partial u}{\partial x}+\frac{\partial v}{\partial y}=0, \tag{1}$$

$$u\frac{\partial u}{\partial x}+v\frac{\partial u}{\partial y}=\frac{\mu_{nf}}{\rho_{nf}}\frac{\partial^2 u}{\partial y^2}-\frac{\sigma_{nf}B^2}{\rho_{nf}}u-\frac{\mu_{nf}}{\rho_{nf}}\frac{u}{k^*}, \tag{2}$$

$$u\frac{\partial T}{\partial x}+v\frac{\partial T}{\partial y}=\frac{\kappa_{nf}}{(\rho C_p)_{nf}}\frac{\partial^2 T}{\partial y^2}-\frac{1}{(\rho C_p)_{nf}}\frac{\partial q_r}{\partial y}+\frac{Q}{(\rho C_p)_{nf}}(T-T_\infty), \tag{3}$$

with proper boundary constraints:

$$at\ y=0: u=u_w(x)\lambda+A_1\frac{\mu_{nf}}{\rho_{nf}}\frac{\partial u}{\partial y}, v=v_w, T=T_w(x)+B_1\frac{\partial T}{\partial y}, \tag{4}$$

$$as\ y\to\infty: u\to 0, T\to T_\infty. \tag{5}$$

Hhere heat generation rate constant $Q = Q_0e^{x/L}$, $A_1 = ae^{-(x/2L)}$, and $B_1 = be^{-(x/2L)}$ indicate, the slip factors following Yan *et al.* [23], for velocity and heat dependence on x, respectively.

On intensifying T^4 in a Taylor series expansion by ignoring higher terms in order, we get,

$T^4 \approx 4T_\infty^3 T-3T_\infty^4$, which can be applied to approximate thermal radiation (Mandal [24])

$$q_r=-\frac{4\sigma^*}{3K^*}\frac{\partial T^4}{\partial y}, \tag{6}$$

Further for the nanofluid, ρ_{nf} the density, σ_{nf} the electrical conductivity, μ_{nf} the coefficient of viscosity, $(\rho Cp)_{nf}$ the capacitance of heat, and κ_{nf} the thermal conductivity, are defined as below (Mandal and Pal [25]):

$$\frac{\rho_{nf}}{\rho_f}=(1-\varphi)+\varphi\frac{\rho_s}{\rho_f}, \frac{\sigma_{nf}}{\sigma_f}=1-\frac{3\varphi\left[1-\frac{\sigma_s}{\sigma_f}\right]}{\left[2+\frac{\sigma_s}{\sigma_f}\right]+\varphi\left[1-\frac{\sigma_s}{\sigma_f}\right]}, \frac{\mu_{nf}}{\mu_f}=\frac{1}{(1-\varphi)^{2.5}},$$

$$\frac{\kappa_{nf}}{\kappa_f}=\frac{\kappa_s+2\kappa_f-2\varphi(\kappa_f-\kappa_s)}{\kappa_s+2\kappa_f+\varphi(\kappa_f-\kappa_s)}, \frac{(\rho C_p)_{nf}}{(\rho C_p)_f}=(1-\varphi)+\varphi\frac{(\rho C_p)_s}{(\rho C_p)_f}$$

We introduced the following similarity transformations (Waini *et al.* [26], Eid and Nafe [27])):

$$\psi = e^{x/2L}\sqrt{(2\nu_f Lc)}f(\eta), \eta = ye^{x/2L}\sqrt{(c/2\nu_f L)}, u = \frac{\partial\psi}{\partial y}, v = -\frac{\partial\psi}{\partial x}, \theta(\eta) = \frac{T-T_\infty}{T_w-T_\infty} \quad \textbf{(7)}$$

By utilizing Eq. (7) as follows, Eq. (1) is trivially satisfied, and Eqs. (2) through (3) are translated towards dimensionless forms:

$$\left(\frac{\mu_{nf}/\mu_f}{\rho_{nf}/\rho_f}\right)f''' + ff'' - 2f'^2 - \left(\frac{\sigma_{nf}/\sigma_f}{\rho_{nf}/\rho_f}\right)Mf' - \left(\frac{\mu_{nf}/\mu_f}{\rho_{hf}/\rho_f}\right)Pf' = 0, \quad \textbf{(8)}$$

$$\frac{1}{Pr}\left[\left\{\frac{\kappa_{nf}}{\kappa_f} + Nr\right\}\theta''\right] + \frac{(\rho C_p)_{nf}}{(\rho C_p)_f}(-f'\theta + f\theta') = 0 \quad \textbf{(9)}$$

These are transformed border constraints:

$$f = S, f' = \lambda + Af'', \theta = 1 + B\theta' \; at \; \eta = 0, \quad \textbf{(10)}$$

$$f' \to 0, \theta \to 0 \; as \; \eta \to \infty . \quad \textbf{(11)}$$

These dimensionally empty parameters are stated as follows: $Pr = \frac{\nu_f(\rho C_p)_f}{\kappa_f}$ is the Prandtl number, $M = \frac{B_0^2\,\sigma_f L}{c\rho_f}$ is the parameter for the magnetic field, $P = \frac{2L\,\nu_f}{c\,\kappa^*}$ is the parameter for the porous field, $Nr = \frac{16\sigma^* T_\infty^3}{3\kappa_f K^*}$ is the parameter for thermal radiation, $\beta = \frac{2\,Q_0 L}{c(\rho C_p)_f}$ is the parameter for heat generation, $B = B_1\sqrt{\frac{c}{2\nu_f L}}$ (Yan *et al.* [23]) is the parameter for thermal slip, $A = A_1\frac{\mu_{nf}}{\rho_{nf}}\sqrt{\frac{c}{2\nu_f L}}$ is the parameter for velocity slip, and $S = -v_0/\sqrt{\nu_f c/2L}$ is the wall mass flux transmit parameter.

Physical Quantities

The physical measures of the material such as significant amounts in flow and heat transmission of nanofluid flow are C_f and Nu_x (skin friction and Nusselt number), which are defined by (Usman *et al.* [28]) are conveyed as:

$$C_f = \frac{\mu_{nf}}{\rho_f u_w^2}\left(\frac{\partial u}{\partial y}\right)_{y=0}, Nu_x = \frac{\left[\kappa_{nf}\left(-\frac{\partial T}{\partial y}\right)_{y=0} + q_r\right]_{y=0}}{\kappa_f(T_w - T_\infty)/2L} = \frac{\left[\kappa_{nf}\left(-\frac{\partial T}{\partial y}\right)_{y=0} + \frac{4\sigma^*}{3K^*}\left(-\frac{\partial T^4}{\partial y}\right)\right]_{y=0}}{\kappa_f(T_w - T_\infty)/2L} \quad \mathbf{(12)}$$

Now, the skin friction coefficient and heat transmit coefficient in the non-dimensional form are given by

$$Re_x{}^{0.5} C_f = \frac{\mu_{nf}}{\mu_f} f''(0), Re_x{}^{-0.5} Nu_x = -\left(\frac{\kappa_{nf}}{\kappa_f} + Nr\right)\theta'(0). \quad \mathbf{(13)}$$

Flow Stability

By incorporating Eqs. (2) and (3) with an unsteady version into Eqs. (14) and (15), respectively, the stability analysis (Yan *et al.* [23]) of the flow is carried out.

$$u\frac{\partial u}{\partial x} + v\frac{\partial u}{\partial y} + \frac{\partial u}{\partial t} = \frac{\mu_{nf}}{\rho_{nf}}\frac{\partial^2 u}{\partial y^2} - \frac{\sigma_{nf}B^2}{\rho_{nf}}u - \frac{\mu_{nf}}{\rho_{nf}}\frac{u}{k^*}, \quad \mathbf{(14)}$$

$$u\frac{\partial T}{\partial x} + v\frac{\partial T}{\partial y} + \frac{\partial T}{\partial t} = \frac{\kappa_{nf}}{(\rho C_p)_{nf}}\frac{\partial^2 T}{\partial y^2} - \frac{1}{(\rho C_p)_{nf}}\frac{\partial q_r}{\partial y} + \frac{Q}{(\rho C_p)_{nf}}(T - T_\infty), \quad \mathbf{(15)}$$

Using the temporal variable $= \left(\frac{c}{2L}\right) te^{x/L}$, a novel similarity transformation is used to achieve the following:

$$\psi = e^{x/2L}\sqrt{(2\nu_f Lc)} f(\eta,\tau), \eta = ye^{x/2L}\sqrt{(c/2\nu_f L)}, u = \frac{\partial\psi}{\partial y}, \theta(\eta,\tau) = \frac{T - T_\infty}{T_w - T_\infty}, \quad \mathbf{(16)}$$

The aforementioned is used to obtain subsequently transformed equations:

$$\left(\frac{\mu_{nf}/\mu_f}{\rho_{nf}/\rho_f}\right)\frac{\partial^3 f}{\partial\eta^3} + f\frac{\partial^2 f}{\partial\eta^2} - 2\left(\frac{\partial f}{\partial\eta}\right)^2 - \frac{\partial^2 f}{\partial\eta\partial\tau} - \left(\frac{\sigma_{nf}/\sigma_f}{\rho_{nf}/\rho_f}\right)M\frac{\partial f}{\partial\eta} - \left(\frac{\mu_{nf}/\mu_f}{\rho_{hf}/\rho_f}\right)P\frac{\partial f}{\partial\eta} = 0, \quad \mathbf{(17)}$$

$$\frac{1}{Pr}\left[\left\{\frac{\kappa_{nf}}{\kappa_f}+Nr\right\}\frac{\partial^2\theta}{\partial\eta^2}\right]+\frac{(\rho C_p)_{nf}}{(\rho C_p)_f}\left(-\theta\frac{\partial f}{\partial\eta}+f\frac{\partial\theta}{\partial\eta}-\frac{\partial\theta}{\partial\tau}\right)+\beta\theta=0 \quad \textbf{(18)}$$

that conditioned to:

$$f(\eta,\tau)=S,\frac{\partial f}{\partial\eta}(\eta,\tau)=\lambda+A\frac{\partial^2 f}{\partial\eta^2}(\eta,\tau),\theta(\eta,\tau)=1+B\frac{\partial\theta}{\partial\eta}(\eta,\tau)\ at\ \eta=0, \quad \textbf{(19)}$$

$$\frac{\partial f}{\partial\eta}(\eta,\tau)\to 0,\theta(\eta,\tau)\to 0\ as\ \eta\to\infty. \quad \textbf{(20)}$$

If the stability of the solutions is examined, the perturbation equations are crucial. By applying stability analysis, $f(\eta)=f_0(\eta)$, $\theta(\eta)=\theta_0(\eta)$, and γ is of eigenvalue set where $\gamma_1<\gamma_2<...<\gamma_{n-1}<\gamma_n$. The perturbation equations are:

$$f(\eta,\tau)=f_0(\eta)+e^{-\gamma\tau}F(\eta),\theta(\eta,\tau)=\theta_0(\eta)+e^{-\gamma\tau}G(\eta), \quad \textbf{(21)}$$

(Weidman *et al.* [3]) where $F(\eta)$, and $G(\eta)$ are comparatively the tiny ones to $f_0(\eta)$, and $\theta_0(\eta)$, accordingly, and is an eigenvalue parameter with an unknown identity that has to be listed. The initial decay or development of the disturbance may be detected by including equation (21) in Eqs. (17 - 20) and τ is set to the value of zero. The resulting linearized eigenvalue issue can then be written as follows:

$$\left(\frac{\mu_{nf}/\mu_f}{\rho_{nf}/\rho_f}\right)F'''+Ff_0''-4f_0'F'+F''f_0+\gamma F'-\left(\frac{\sigma_{nf}/\sigma_f}{\rho_{nf}/\rho_f}\right)MF'-\left(\frac{\mu_{nf}/\mu_f}{\rho_{hf}/\rho_f}\right)PF'=0, \quad \textbf{(22)}$$

$$\frac{1}{Pr}\left[\left\{\frac{\kappa_{nf}}{\kappa_f}+Nr\right\}G''\right]+\frac{(\rho C_p)_{nf}}{(\rho C_p)_f}(F\theta_0'+G'f_0-\theta_0F'-Gf_0'+\gamma G)+\beta G=0, \quad \textbf{(23)}$$

that was trained under linear circumstances:

$$F(0)=0,\ F'(0)=AF''(0),G(0)=BG'(0), \quad \textbf{(24)}$$

$$F'(\infty)\to 0,G(\infty)\to 0. \quad \textbf{(25)}$$

In order to obtain a nontrivial solution (eigenvalue), Harris *et al.* [29] recovered boundary constraints with a new constraint to offer a sufficient range of eigenvalues. Thus, $F'(\infty)\to 0$ is retrieved in this research with $F''(0)=1$. A positive value of the resultant lowest eigenvalue γ_1 ($\gamma_1>0$) indicates a solution that is physically feasible and realistic, whereas γ_1 with a negative value denotes an impractical solution.

RESULTS AND DISCUSSION

The physical properties of nanofluids are taken from the literature review of Mandal and Pal [30] and Giri *et al.* [31] (see Table **1**). The numerical outcomes are analysed to explore and visualize the physical impact of the governing variables to identify the key flow with the heat transport behavior of the magneto-radiative *Ag*-water nanofluid across a permeable medium over an exponential contracting surface in the presence of heat source, velocity, and thermal slip boundary constraints. The significance of calculated findings is discussed concerning the current issue of heat transmit characteristics, and the dual solutions in the nanofluid are examined in this part. Numerical computing was made easier by Matlab's platform's available bvp4c function. The dual findings in the solution have also received a lot of attention in this work. As a result, two sets of guesses at the starting and the proper thickness, $\eta_\infty = 15$ at the boundary layer, are offered to perceive the dual solutions with the concern of the pertinent parameter values.

Table 1. Physical properties (Mandal and Pal [30], Giri *et al.* [31]).

Nanoparticles and base fluid(water)'s properties					
Properties	Cp	ρ	κ	σ	Pr
(Units)	$(Jkg^{-1}K)$	$(kg\ m^{-3})$	$(\mathrm{Wm^{-1}\ K})$	$(\Omega \mathrm{m}^{-1})$	
Ag	235	10490	4029	6.3×10^7	--
Water	4179	997.1	0.613	0.05	6.2

We compared our results with previously published findings of Ghosh & Mukhopadhyay [5] and Waini *et al.* [26] for specified values of parameters, $\phi = M = Nr = P = A = B = \beta = 0$, $S = 3.0$, $Pr = 0.7$, with $\lambda = -1$ in Table **2**. We discovered a reasonable connection between the current and previously reported findings. As a result, the results presented in this study are guaranteed by this approach. As a result, the procedure is both legitimate and acceptable. Table **3** shows the shear stress (skin friction coefficient) and heat transmit rate (Nusselt number) computed for the first solution (upper branch) and second solution (lower branch) as a result of changing parameters when $\lambda = -2.0$, and $Pr = 6.2$ for *Ag*-water nanofluid. According to the data, the skin friction factor $(Re_x^{1/2}C_f)$ considerably increases for the stable first solution branch with ϕ, M, P, and S but diminishes with A, respectively. The computed Nusselt number values ($Re_x^{-1/2} Nu_x$) for the first solution branch solution show a considerable increase with ϕ, M, P, A, S, and β but diminish for B, respectively. Also, In Figs. (**2**) through (Fig. **6**), the analysis of particular parameters related to the $Re_x^{1/2}\, C_f$ (skin-friction) and $Re_x^{-1/2}\, Nu_x$ (Nusselt number), with f'(η) (velocity) and $\theta(\eta)$ (temperature) profiles plots for *Ag*-water nanofluid, are shown. For a certain range of the contracting variable λ, which relies on the suction parameter's strength, all the figures show the presence of multiple (dual) solutions, satisfying the given boundary constraints. There are unique outcomes for $\lambda = \lambda_c$, no solution presents for the range of $\lambda < \lambda_c$, and several outcomes present in the range of $\lambda > \lambda_c$. At this bifurcation point $\lambda = \lambda_c$, all the approximations for the boundary layer fail; hence, we cannot find any solution in the range of $\lambda < \lambda_c$.

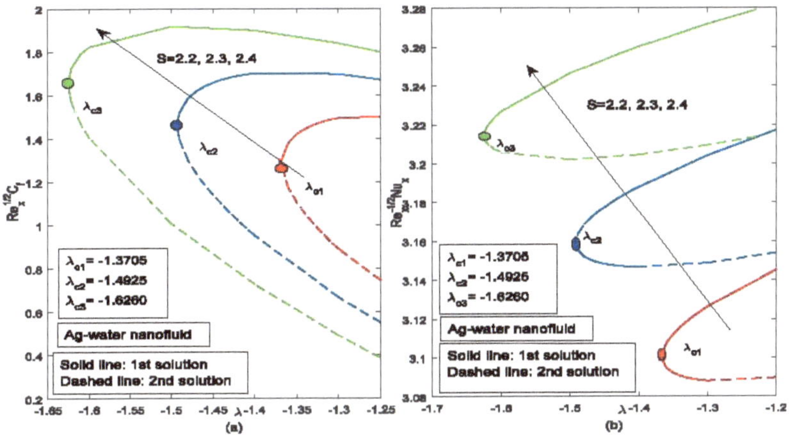

Fig. (2). Skin friction coefficient and Nusselt number with λ for varied S with $\phi = 0.01$, $M = 0.1$, $P = 0.1$, $Nr = 0.1$, $A = 0.2$, $B = 0.2$, $\beta = 0.1$, and $Pr = 6.2$.

Table 2. Comparison table for viscous fluid ($\phi = 0$) when Pr = 0.7, S = 3, and M =P = Nr = B = A = β = 0 for the contracting surface ($\lambda = -1$).

-	f''(0)		-θ'(0)	
-	1st solution	2nd solution	1st solution	2nd solution
Ghosh & Mukhopadhyay [5]	2.39082	−0.97223	1.7712	0.84832
Waini *et al.* [26]	2.390814	−0.972247	1.771237	0.848316
Prsent Results	2.3908106	−0.9721278	1.7712345	0.8478432

Table 3. Calculated value of skin-friction coefficient ($Re_x^{1/2}C_f$) and Nusselt number ($Re_x^{-1/2}Nu_x$) for nanofluid (*Ag-water*) with $\lambda = -2.0$, and $Pr = 6.2$.

ϕ	Nr	M	P	A	B	S	β	$Re_x^{1/2}C_f$ (1st *solution*)	$Re_x^{1/2}Nu_x$ (2nd *solution*)	$Re_x^{1/2}C_f$ (1st *solution*)	$Re_x^{1/2}Nu_x$ (2nd *solution*)
0.10	0.5	0.10	0.10	0.1	0.1	2.4	0.1	4.308586881	2.975166047	7.169450021	6.970105137
0.11								4.731809822	2.819208543	7.225288181	6.932116992
	0.6							4.731809822	2.819209615	7.326706928	6.992652200
		0.11						4.756839775	2.785978416	7.330145083	6.985087131
			0.11					4.780509171	2.754298142	7.333380157	6.977792759
				0.2				4.679941861	1.868307418	7.605921450	6.945653678
					0.2			4.679941861	1.868307435	5.487794513	5.135553197
						2.5		4.971040968	1.456575776	5.634120222	5.246753061
							0.2	4.971040968	1.456576429	5.614756778	5.205332317

In the present work, water (H_2O) is used as the base fluid, and silver(*Ag*) is used as the nanoparticle to create the *Ag*-water nanofluid. The sheets' contracting ($\lambda < 0$) surfaces are taken into account. It is important to note that the *Pr* (= 6.2) remained constant throughout the whole investigation. The variation of the skin friction coefficient ($Re_x^{1/2} C_f$) and Nusselt number ($Re_x^{-1/2}Nu_x$) *versus* the contracting surface parameter is depicted in Figs. (**2a** through **2b**) for some reasonable values of the parameter S (= 2.2, 2.3, and 2.4) for = 0.01, respectively. As shown in the figures, dual solutions are present for S (= 2.2, 2.3, 2.4) when the critical values of λ (λ_c), are $\lambda_{c1} = -1.3705$, $\lambda_{c2} = -1.4925$, and $\lambda_{c3} = -1.6260$, respectively. Furthermore, it shows that the curves are shifting from the higher right to the lower left of the domain, which suggests that the skin friction coefficient ($Re_x^{1/2} C_f$) and Nusselt number ($Re_x^{-1/2} Nu_x$) can be increased when the contracting parameter (λ) magnifies. According to Figs. (**2a-2b**), we discovered that when the suction parameter accelerates, the skin-friction value with Nusselt number tends to increase for the solution branches for *Ag*-water nanofluid. Also, Figs. (**3a-3b**) illustrate the impacts of the volume fraction ϕ(=0.001, 0.05, 0.1) of silver nanoparticles over the contraction parameter λ, skin friction ($Re_x^{1/2} C_f$) and the Nusselt number ($Re_x^{-1/2} Nu_x$). The presence of multiple solutions is evident in Figs. (**3a** and **3b**) for ϕ (=0.01, 0.05, 0.1), when $\lambda_{c1} = -1.6260$, $\lambda_{c2} = -1.9953$, and $\lambda_{c3} = -2.3500$, respectively. It shows that the boundary layer's critical point presents at the area of the contracting surface and the existence of the silver nanoparticle causes a delay in the boundary layer's bifurcation. Both skin friction and Nusselt number at the contracting sheet area for the 1st and 2nd solutions increase with the upsurge in the silver nanoparticle fraction parameter ϕ.

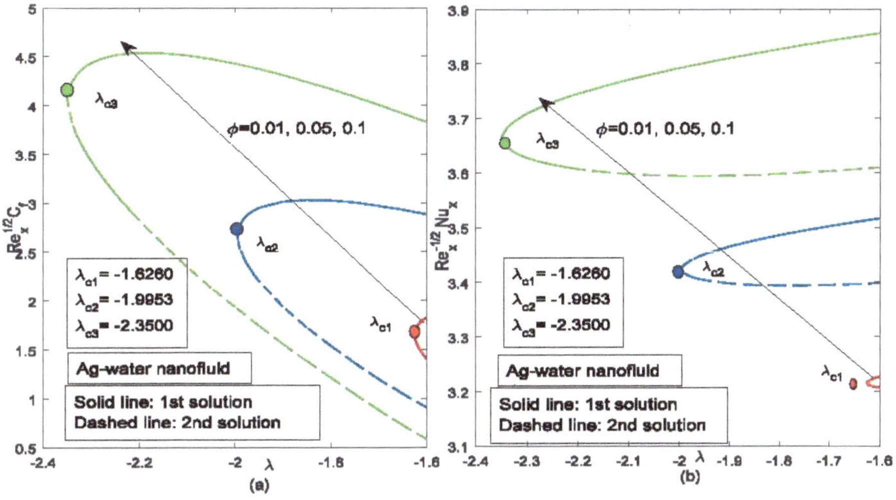

Fig. (3). Skin friction coefficient and Nusselt number with λ for varied ϕ with $S = 2.4$, $M = 0.1$, $Nr = 0.1$, $P = 0.1$, $A = 0.1$, $B = 0.1$, $\beta = 0.1$, and $Pr = 6.2$.

The impacts of various magnetic parameter M (= 0.0, 0.05, and 0.1) values on the velocity parameter for *Ag*-water nanofluid are shown in Fig. (**4a**). It demonstrates that as M is raised, the hydrodynamic velocity rises in the first zone of solutions and falls in the second, respectively. Physically, the advection of the fluid velocity causes the reverse flow across the sheet to upsurge with a hike in magnetic parameter. The velocity lines for *Ag*-water over a contracting sheet in Fig. (**4b**) represent different values of the porosity variable P (= 0.0, 0.05, and 0.1). Due to the lift in P, the velocity field rises for the 1st solution in this Fig. while falls for the 2nd solution. The nanofluid's velocity is physically boosted by including a permeable component in an original solution that forces the fluid to be dragged toward the fluid motion. The temperature profile of the *Ag*-water is shown in Fig. (**5a**), along with how both solutions (first and second) increase when the thermal emission parameter Nr (= 0.5, 1.5, and 3.0) increases. Based on this graph, we assert that an upper estimate of Nr causes the temperature of the nanofluid to increase more quickly. The physical justification for this is that higher thermal radiation implies the occurrence of Rosseland radiative absorptive K^* diminutions , which cause the radiative heat flux to increase the thermal boundary thickness. Fig. (**5b**) shows the effects of heat production parameter β (= 0.0, 0.1, and 0.2) on the temperature field of the nanofluid (*Ag*-water). These data of the nanofluids (*Ag*-water) show that the temperature line rises when the heat production parameter β increases. If heat generation is regulated, it is possible to achieve the necessary heat transmit performance.

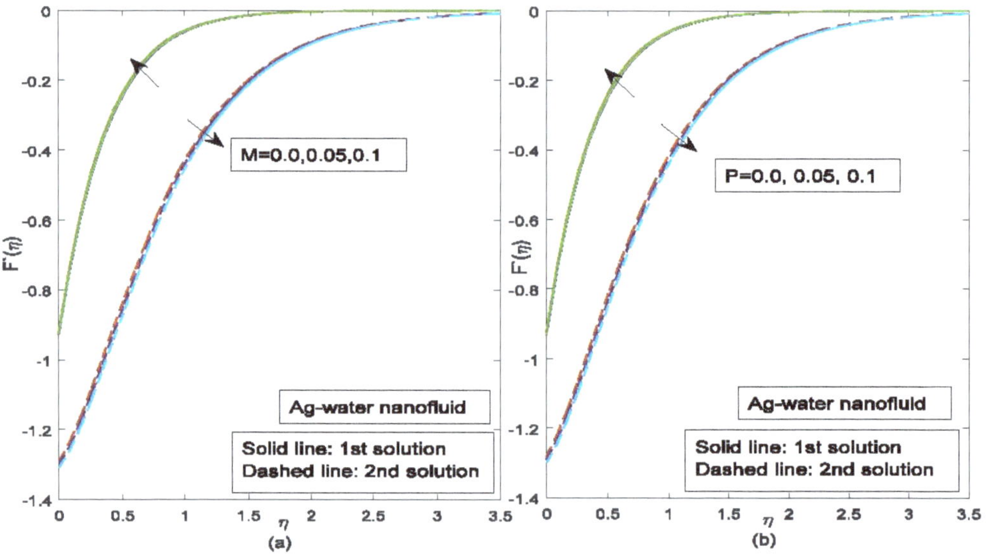

Fig. (4). Velocity profiles for various values of M and P with $\phi = 0.1$, $Nr = 0.5$, $S = 2.2$, $A = 0.2$, $B = 0.2$, $\beta = 0.1$, $\lambda = -1.4$, and $Pr = 6.2$.

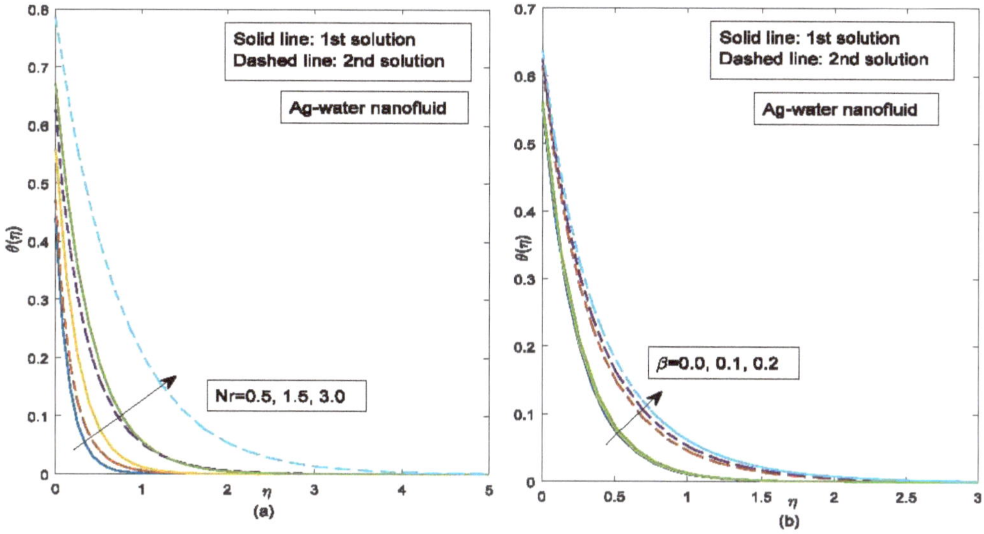

Fig. (5). Temperature profiles for various values of Nr and β with $\phi = 0.1$, $M = 0.05$, $P = 0.05$, $S = 2.4$, $A = 0.2$, $B = 0.2$, $\beta = 0.1$, $\lambda = -1.4$ and $Pr = 6.2$.

The velocity profiles for *Ag*-water nanofluid with increments of the velocity slip variable A (= 0.2, 0.4, and 0.6) are portrayed in Fig. (**6a**). It is seen that while the velocity line decreases for the 2nd solution, it increases for the 1st solution. Physically, when the sheet contracts, the flow resistance decreases, allowing more fluid to pass along the sheet and increasing the velocity of the flow. As a result, increasing the flow velocity requires a high value of velocity slip, and *vice versa*. As shown in Fig. (**6b**), the *Ag*-water nanofluid has less effect on the temperature profile due to the thermal slip parameter B. Physically, raising the slip parameter results in reduced surface friction, which, in turn, results in a reduction in fluid temperature. Moreover, for all the figures, the boundary layer width of the top branch of the first primary solution is thinner than that of the second next solution (lower branch). Also, the asymptotic fulfillment of the infinite boundary constraints confirms the accuracy of the calculated findings. Fig. (**7**) shows the streamlines for the first with the second solutions for $\lambda = -1.5$, $\phi = 0.1$, P=0.05, Nr= 0.5, $M = 0.05$ $A = 0.1$, $B = 0.1$, β= 0.1 and $Pr = 6.2$, respectively. With the first solution, the streamlined pattern is straightforward, but for the second, we can see that the boundary layer thickness from the starting point is greater than for the first. In addition, we carefully looked at the flow stability brought on by the emergence of dual solutions. We made the numerical procedure's bvp4c solver easier. The minimal eigenvalues γ_1 are useful in indicating the stability of the flow phenomena by using this method. For a few favored values of ϕ, λ, and S, with P=0.1, Nr= 0.5, $M = 0.1$, $A = 0.1$, β= 0.1, $B = 0.1$, and $Pr = 6.2$, Table **4** shows the estimated lowest eigenvalue for both solution branches. This table reveals that if the values of ϕ, λ, and S decrease, the smallest eigenvalues in the 1st and 2nd solution branches converge with each other. It is also interesting to note that the 1st solution branch results in positive smallest eigenvalues while the 2nd solution branch results in negative smallest eigenvalues. It is also mentioned that when the crucial point λ_c got close, the smallest eigenvalues started to approach zero. After solving Eqs. (22)-(23) with BCs (24)-(26), the smallest or minimal eigenvalues γ_1 are displayed in Fig. (**8**). The lowest eigenvalues for two different values of S (= 2.3, 2.4) are displayed in this image concerning the contracting sheet parameter λ with critical values of $\lambda_{c1} = -2.4900$ for $S = 2.3$ and $\lambda_{c2} = -2.750$ for $S = 2.4$, respectively. It is interesting to observe that there is no smallest eigenvalue for $\lambda < \lambda_c$. The stable solution zone is represented by the top half of Fig. (**8**), while the unstable solution region is represented by the lower half. It is noticed that $e^{\gamma_1 \tau} \rightarrow 0$ as time progresses ($\tau \rightarrow \infty$) for positive values of γ_1. On the other hand, negative γ_1 values, $e^{\gamma_1 \tau} \rightarrow \infty$ .These findings also support the first solution's stability and long-term physical dependability, and *vice versa*.

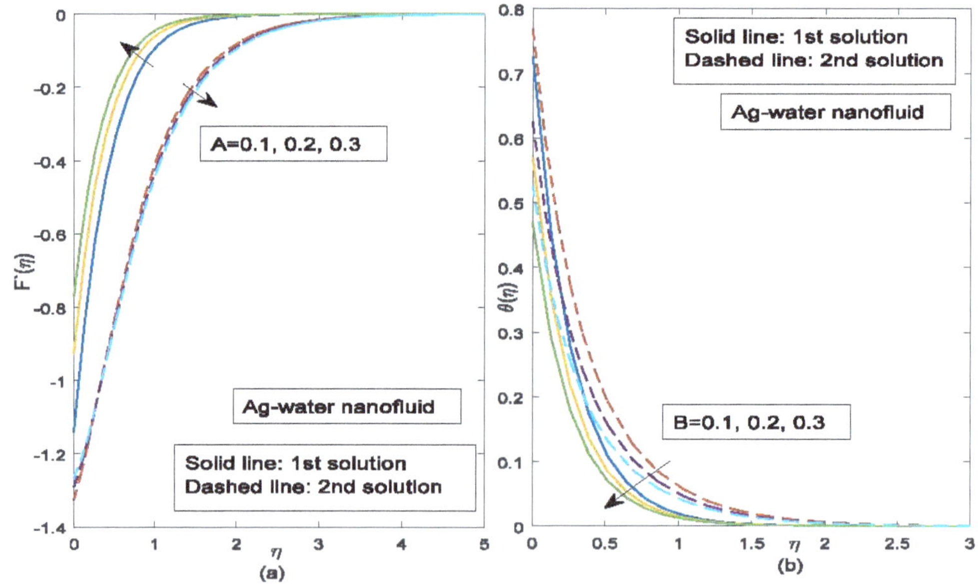

Fig. (6). Velocity and temperature profiles for various values of A and B with $\phi = 0.1$, $M = 0.05$, $Nr = 1.5$, $S = 2.2$, $\beta = 0.1$, $\lambda = -1.4$, and $Pr = 6.2$.

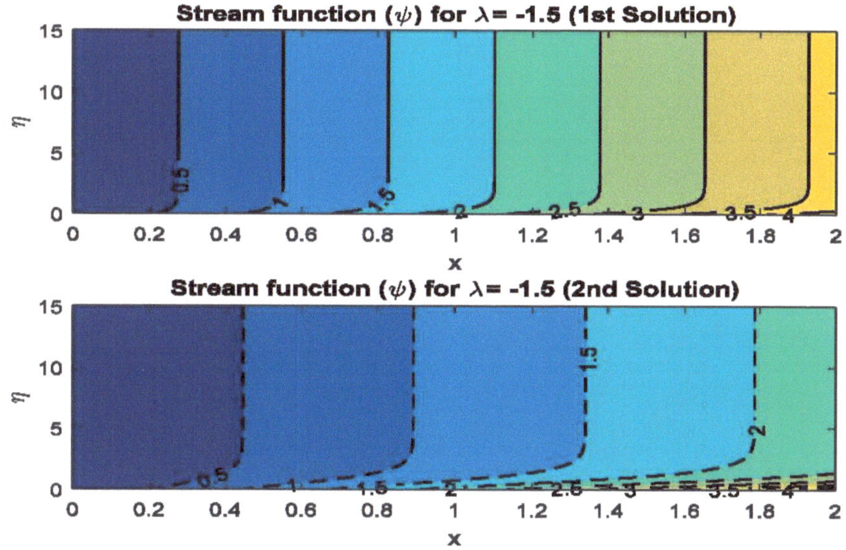

Fig. (7). Streamline flow for $\phi = 0.1, M = 0.05$, Nr=0.5, $P = 0.05$, $S = 2.4, A = 0.2, B = 0.2$, $\beta = 0.1$, $\lambda = -$and $Pr = 6.2$.

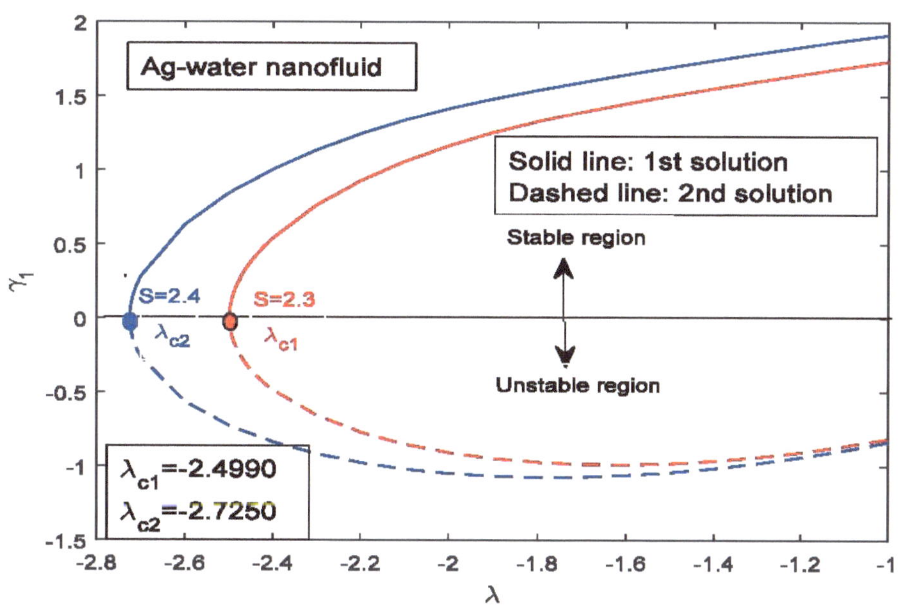

Fig. (8). Smallest eigenvalue with λ with different S for $\phi = 0.1$, $M = 0.1$, Nr=0.5, $P = 0.1$, $A = 0.1$, $B = 0.1$, $\beta = 0.1$, and $Pr = 6.2$.

Table 4. The least eigenvalues with 1st and 2nd solutions for the most desired values of ϕ, λ, and S as P=0.1, M=0.1, Pr = 6.2, Nr= 0.5, B = 0.1, A = 0.1, and β= 0.1.

Least Eigen Values(γ_1)				
ϕ	A	S	1st solution	2nd solution
0.01	-1.7	2.3	1.156857382	−0.430899783
0.05			1.237286326	−0.812757762
0.10			1.390883937	−0.986223539
-	-1.8		1.327345880	−0.973248156
	-1.9		1.402882493	−0.948051346
	-2.0		1.670107497	−0.908597512
	-	2.4	1.643990609	−1.048636632
		2.3	1.670107497	−0.908597512
		2.2	1.444716711	−0.719819501

CONCLUSION

This article's goal was to look at stability analysis for dual solutions that were obtained for the radiative magnetic flow of *Ag*-water nanofluid *via* a permeable medium across a permeable exponentially shrinkable surface in the existence of heat generation and experiencing velocity-thermal slip. The flow and transmission of energy in the nanofluid are influenced by several physical parameters with a mild volume concentration of silver (*Ag*) nanoparticles in water. The numerical results for the lower with upper branch solutions were generated using Bvp4c in MATLAB. Investigations were also conducted into the impression of the friction factor and the rate of heat transmission at the surface's wall surface. The results are as follows:

(i) The silver (Ag) nanoparticles in *Ag*-water nanofluid enhance the heat transmit performance of the system. The Nusselt number for the 1st solution might be increased by improving the suction parameters and increasing the volume percentage of silver nanoparticles, whereas the tendency is the opposite for second-branch solutions. For both branches of solutions, the contracting parameter increment increases the Nusselt number.

(*ii*) The fluid velocity increases for the 1st solution when the porosity, magnetic field, and velocity slip parameters are increased, whereas it decreases for the 2nd solution in the region of the contracting surface.

(*iii*) In the region of the contracting surface, the temperature profile can increase the thermal emission parameter and the heat production parameter, but it decreases with the progression of thermal slippage for both the 1st and 2nd solutions.

(*iv*) Only the 1st of the two solutions is stable because only the 1st solution is possible in the presence of a contracting surface and the right amount of suction. Positive eigenvalues are formulated by the 1st solution branch, but negative eigenvalues are constructed by the 2nd solution. Both smallest eigenvalues converge to zero by reducing the values of silver nanoparticle volume fraction, suction, and contracting parameter parameters.

(*v*) The current study may be further expanded to address issues including boundary layer movement with variable conductivity and viscosity, mixed convection, mass transmission, chemical reactions, micropolar nanofluid, and many other cutting-edge aspects.

CONFLICT OF INTEREST

The author certifies that they have no financial or personal conflicts of interest or known personal ties that would have seemed to have an impact on the work disclosed in this publication.

REFERENCES

[1] S.U.S. Choi, and J.A. Eastman, *Enhancing Thermal Conductivity of Fluids with Nanoparticles.* Department of Energy: Washington, DC, USA, 1995.

[2] J.H. Merkin, "On dual solutions occurring in mixed convection in a porous medium", *J. Eng. Math.*, vol. 20, no. 2, pp. 171-179, 1986.
http://dx.doi.org/10.1007/BF00042775

[3] P.D. Weidman, D.G. Kubitschek, and A.M.J. Davis, "The effect of transpiration on self-similar boundary layer flow over moving surfaces", *Int. J. Eng. Sci.,* vol. 44, no. 11-12, pp. 730-737, 2006.
http://dx.doi.org/10.1016/j.ijengsci.2006.04.005

[4] S. Dero, M.J. Uddin, and A.M. Rohni, "Stefan blowing and slip effects on unsteady nanofluid transport past a contracting sheet: Multiple solutions", *Heat Transf. Asian Res.*, vol. 48, no. 6, pp. 2047-2066, 2019.

http://dx.doi.org/10.1002/htj.21470

[5] S. Ghosh, and S. Mukhopadhyay, "Stability analysis for model-based study of nanofluid flow over an exponentially shrinking permeable sheet in presence of slip", *Neural Comput. Appl.*, vol. 32, no. 11, pp. 7201-7211, 2020.

http://dx.doi.org/10.1007/s00521-019-04221-w

[6] M.M. Bhatti, T. Abbas, and M.M. Rashidi, "Numerical study of entropy generation with nonlinear thermal radiation on magnetohydrodynamics non-Newtonian nanofluid through a porous contracting sheet", *J. Magn.*, vol. 21, no. 3, pp. 468-475, 2016.

http://dx.doi.org/10.4283/JMAG.2016.21.3.468

[7] A.K. Tiwari, F. Raza, and J. Akhtar, *MHD flow with Silver (Ag) and Titanium (TiO_2) water based Nanofluid over a non-linearly extending/contracting sheet.* 2017.

[8] H.B. Lanjwani, M.S. Chandio, K. Malik, and M.M. Shaikh, "Stability analysis of boundary layer flow and heat Transf. of Fe_2O_3 and Fe-water base nanofluid over a extending/contracting sheet with radiation effect", *Eng. Technol. Appl. Sci. Res.*, vol. 12, no. 1, pp. 8114-8122, 2022.

http://dx.doi.org/10.48084/etasr.4649

[9] H.B. Lanjwani, M.I. Anwar, A. Wahab, S.A. Shehzad, and M. Arshad, "Analysis of triple solutions in mixed convection flow and heat transfer characteristics of Ag-water based nanofluid over porous shrinking/stretching sheet", *Mater. Sci. Eng. B*, vol. 286, p. 116076, 2022.

http://dx.doi.org/10.1016/j.mseb.2022.116076

[10] G. Rasool, X. Wang, U. Yashkun, L.A. Lund, and H. Shahzad, "Numerical treatment of hybrid water based nanofluid flow with effect of dissipation and Joule heating over a shrinking surface: Stability analysis", *J. Magn. Magn. Mater.*, vol. 571, p. 170587, 2023.

http://dx.doi.org/10.1016/j.jmmm.2023.170587

[11] G. Mandal, and D. Pal, "Stability analysis of radiative-magnetic hybrid nanofluid slip flow due to an exponentially extending/contracting permeable sheet with heat generation", *Int. J. of Ambient Energy*, vol. 0, pp. 1-12, 2023.

[12] S. Chaudhary, and K.M. Kanika, "Heat generation/absorption and radiation effects on hydromagnetic stagnation point flow of nanofluids toward a heated porous extending/contracting sheet with suction/injection", *J. of Porous Media*, vol. 23, vol. 1, pp.27-49, 2020.

[13] N.C. Roy, and I. Pop, "Unsteady magnetohydrodynamic stagnation point flow of a nanofluid past a permeable shrinking sheet", *Zhongguo Wuli Xuekan*, vol. 75, pp. 109-119, 2022.

http://dx.doi.org/10.1016/j.cjph.2021.12.018

[14] B. Ishtiaq, and S. Nadeem, "Theoretical analysis of Casson nanofluid over a vertical exponentially shrinking sheet with inclined magnetic field", *Waves Random Complex Media*, vol. 0, pp. 1-17, 2022.

http://dx.doi.org/10.1080/17455030.2022.2103206

[15] R.M. Muntazir, M. Mushtaq, S. Shahzadi, and K. Jabeen, "MHD nanofluid flow around a permeable extending sheet with thermal radiation and viscous dissipation", Proceedings of the Institution of Mechanical Engineers, Part C", *J. Mech. Eng. Sci.,* vol. 236, no. 1, pp. 137-152, 2022.

http://dx.doi.org/10.1177/09544062211023094

[16] G. Mandal, and D. Pal, "Entropy analysis of magneto-radiative SWCNT-MWCNT/H 2 O hybrid nanofluid flow with slip boundary conditions", *Int. J. Amb. Energy,* vol. 44, no. 1, pp. 1017-1030, 2023.

http://dx.doi.org/10.1080/01430750.2022.2161631

[17] D. Pal, and G. Mandal, "Thermal radiation and MHD effects on boundary layer flow of micropolar nanofluid past a stretching sheet with non-uniform heat source/sink", *Int. J. Mech. Sci.,* vol. 126, pp. 308-318, 2017.

http://dx.doi.org/10.1016/j.ijmecsci.2016.12.023

[18] Norfifah Bachok, S.S.P.M. Isa, N.M. Arifin, and N. Bachok, "Thermal radiation in nanofluid penetrable flow bounded with partial slip condition", *CFD Letters,* vol. 13, no. 8, pp. 32-44, 2021.

http://dx.doi.org/10.37934/cfdl.13.8.3244

[19] M. Azam, "Effects of Cattaneo-Christov heat flux and nonlinear thermal radiation on MHD Maxwell nanofluid with Arrhenius activation energy", *Case Stud. Therm. Eng.,* vol. 34, p. 102048, 2022.

http://dx.doi.org/10.1016/j.csite.2022.102048

[20] U. Farooq, H. Waqas, T. Muhammad, M. Imran, and A.S. Alshomrani, "Computation of nonlinear thermal radiation in magnetized nanofluid flow with entropy generation", *Appl. Math. Comput.,* vol. 423, p. 126900, 2022.

http://dx.doi.org/10.1016/j.amc.2021.126900

[21] H.A. Alzahrani, A. Alsaiari, J.K. Madhukesh, R.N. Kumar, and B.M. Prasanna, "Effect of thermal radiation on heat Transf. in plane wall jet flow of Casson nanofluid with suction subject to a slip boundary condition", *Waves Random Complex Media,* vol. 0, pp. 1-18, 2022.

[22] N.S. Wahid, N.M. Arifin, N.S. Khashi'ie, and I. Pop, "Hybrid nanofluid slip flow over an exponentially extending/contracting permeable sheet with heat generation", *Mathematics,* vol. 9, no. 1, p. 30, 2020.

http://dx.doi.org/10.3390/math9010030

[23] L. Yan, S. Dero, I. Khan, I.A. Mari, D. Baleanu, K.S. Nisar, E.S.M. Sherif, and H.S. Abdo, "Dual solutions and stability analysis of magnetized hybrid nanofluid with joule heating and multiple slip conditions", *Processes,* vol. 8, no. 3, p. 332, 2020.

http://dx.doi.org/10.3390/pr8030332

[24] G. Mandal, "Convective radiative heat Transf. of micropolar nanofluid over a vertical non-linear extending sheet", *J. Nanofluids,* vol. 5, no. 6, pp. 852-860, 2016.

http://dx.doi.org/10.1166/jon.2016.1265

[25] G. Mandal, and D. Pal, "Dual solutions for magnetic-convective-quadratic radiative MoS_2–SiO_2/H_2O hybrid nanofluid flow in Darcy-Fochheimer porous medium in presence of

second-order slip velocity through a permeable shrinking surface: Entropy and stability analysis", *Int. J. Model. Simul.*, pp.1-27, 2023.

[26] I. Waini, A. Ishak, and I. Pop, *Multiple solutions of the unsteady hybrid nanofluid flow over a rotating disk with stability analysis.,* vol. 94. European J. of Mechanics-B/Fluids, 2022, pp. 121-127.

[27] M.R. Eid, and M.A. Nafe, "Thermal conductivity variation and heat generation effects on magneto-hybrid nanofluid flow in a porous medium with slip condition", *Waves Random Complex Media,* vol. 32, no. 3, pp. 1103-1127, 2022.

http://dx.doi.org/10.1080/17455030.2020.1810365

[28] M. Usman, M. Hamid, T. Zubair, R. Ul Haq, and W. Wang, "Cu-AlO/Water hybrid nanofluid through a permeable surface in the presence of nonlinear radiation and variable thermal conductivity *via* LSM", *Int. J. Heat Mass Transf.,* vol. 126, pp. 1347-1356, 2018.

http://dx.doi.org/10.1016/j.ijheatmasstransfer.2018.06.005

[29] S.D. Harris, D.B. Ingham, and I. Pop, "Mixed convection boundary layer flow near the stagnation point on a vertical surface in a porous medium: Brinkman model with slip", *Transp. Porous Media,* vol. 77, no. 2, pp. 267-285, 2009.

http://dx.doi.org/10.1007/s11242-008-9309-6

[30] G. Mandal, and D. Pal, "Dual solutions of radiative $Ag - MoS_2$/water hybrid nanofluid flow with variable viscosity and variable thermal conductivity along an exponentially contracting permeable Riga surface: Stability and entropy generation analysis", *Int. J. of Modelling and Simulation,* pp. 1-26, 2023.

[31] S.S. Giri, K. Das, and P.K. Kundu, "Computational analysis of thermal and mass transmit in a hydromagnetic hybrid nanofluid flow over a slippery curved surface", *Int. J. Ambi. Energy,* vol. 43, no. 1, pp. 6062-6070, 2022.

http://dx.doi.org/10.1080/01430750.2021.2000491

CHAPTER 7

Spectral Quasi-Linearization Method for Thixotropic Nanofluid Passing through a Stretching Surface with Activation Energy

Anwesha Dingal[1], Puspita Mondal[1], Anindya Kundu[1], Sharmistha Ghosh[1] and Hiranmoy Mondal[1,*]

[1] *Department of Applied Mathematics, Maulana Abul Kalam Azad University of Technology, Haringhata, Nadia, West Bengal, India*

Abstract: The primary goal of the current investigation is to understand the thixotropic nanofluid's magnetohydrodynamic (MHD) nonlinear convective flow. The implementation of appropriate transformations allows for the transition of sets of ordinary differential equations from partial differential systems. Spectral Quasi-Linearization Method (SQLM) is applied to solve these ODEs. Further study has been performed in order to establish a model for nanomaterials containing thermophoresis phenomena and Brownian motion. Graphs are used to depict and discuss the effects caused due to numerous quantities of fluid flow, temperature, environment, and nanoparticle concentration. In addition to this, the local Nusselt number, the skin friction coefficient, as well as the Sherwood number's numerical values, are determined and examined. Some interesting observations on velocity, heat, and concentration have been observed due to the variation made in thixotropic parameters and other allied parameters such as thermophoresis and Brownian motion.

Keywords: Newtonian heat and mass conditions, Magnetohydrodynamics (MHD), Spectral quasilinearization method (SQLM), Thixotropic nanofluid.

INTRODUCTION

There are currently a large number of applications for non-Newtonian materials in technology and business. Due to numerous applications in physiology, medicines, oil reservoirs, coating of wires, grease, fibre technology, physiology, food products (such as crystal formation, milk, and apple sauce), *etc.*, researchers and scientists have been motivated to consider this. Fetecau *et al.* [1] examined the energy balance for an Oldroyd-B fluid flowing through a straight plate under a time-varying shear stress. Wang and Tan WC [2] analysed the stability of a convective double-diffusive Maxwell fluid in a porous medium. Despite these challenges, most

*Corresponding author Hiranmoy Mondal:** Department of Applied Mathematics, Maulana Abul Kalam Azad University of Technology, Haringhata, Nadia, West Bengal; E-mail: hiranmoymondal@yahoo.co.in

researchers are always attempting to analyze the behaviour of different non-Newtonian fluids from numerous perspectives. Shear thinning and thixotropic fluid vary in such a way that the latter demonstrates viscosity reduction in a liquid which is shear thinning in nature with an increase in shear rate and the former indicates viscosity degradation over time. One can examine a few studies on thixotropic substances. Qayyum and Hayat [3] examined the MHD nonlinear thixotropic nanofluid flow under Newtonian heat, mass conditions, and chemical reactions. Olubode and Nehad [4] explored the examination of MHD thixotropic nanofluid bioconvection flow past a vertical surface in the presence of gyrotactic bacteria and nanoparticles. Mehta and Chouhan [5] examined the MHD flow of nanofluids through a vertical channel while being exposed to radiation, heat, and porous media. Nanofluid is a word used to describe the diluted suspension of fibres and nanometer-sized particles in a fluid. In comparison to typical fluids, the mixing of nanoparticles altered the physical characteristics of the liquid, such as density, thermal conductivity, viscosity, *etc.* Dharmaiah, and Dinarv [6] analysed the Buongiorno non-homogeneous two-component model in the presence of activation energy for Howarth's wavy cylinder. Shafiq *et al.* [7] used a machine learning algorithm to establish the implication of bioconvective flow of MHD thixotropic nanofluid traversing a vertical surface. Waqas *et al.* [8] analysed the joule heating impacts and nonlinear convection in the flow of magneto-thixotropic nanofluid under convection by a heated variable over a thicked surface. The use of nanoparticles is widespread in a variety of fields, including smart computers, renewable energy, solar cells, catalysis, materials, electronics, optics, and many more. Currently, there are two categories in which the idea of energy transmission in nanofluids might be placed. One category of nanofluids was assumed to be composite, composed of nanoparticles, and shielded by nanolayers. Anantha [9] analysed MHD nonlinear radiative slip motion across a stretching sheet of a non-Newtonian fluid numerically examined in a porous medium. Madiha, *et al.* [10] numerically analysed the unsteadiness of the three-dimensional Williamson fluid-particle flow when suspended under MHD thermal radiations. Kataria *et al.* [11] established the impact of nonlinear radiation on fluid flow for MHD with optimised entropy. The studies contain some inquiries into this idea. According to the alternative idea, micro-mixing results from a combination of particle convention and Brownian motion, which determines how well a nanofluid conducts heat. In these models, particle dynamics is taken into account. Through the use of this theory, the particle interaction in nanofluids can be seen. In general, four categories of heating mechanisms define the boundary temperature fields. These categories include (i) arbitrary or constant heat flux on the surface; (ii) conjugate conditions; and (iii) arbitrary or constant wall temperature, in which temperature has been moved to a convective fluid passing through a boundary surface along with a limited specific heat capacity and thickness. The merging temperature is determined by the underlying assumptions of the system, specifically the thermal conductivity of solid and liquid, which is unknown beforehand. Conjugate convective flow is the primary characteristic of Newtonian heating and the rate of heat transfer from the boundary surface, which has a restricted specific heat, is related to the local surface temperature. Heat exchangers, solar radiation, storage of thermal energy, the petroleum sector, and other fields are among Newtonian heating's applications. Das [12] established in a rotating reference frame, that there is an impact of thermal radiation and chemical reaction over the mass and heat transfer of MHD micropolar fluid flow. Jamil [13] established fractional derivatives with heat radiation and chemical reactions which are used to represent MHD Maxwell flow.

In this article, the convective flow of a non-linear thixotropic nanofluid is examined in the presence of a magnetic field, Newtonian heat, and mass conditions. This paper is extended in which the flow of thixotropic nanofluid passing through a stretching sheet is examined under activation energy and thermal radiation. Here spectral quasi-linearization method is used. This Brownian motion, heat absorption/generation, and thermophoresis are studied. Quasilinearisation is used to linearize nonlinear ordinary differential equations. The use of a spectral method in order to solve the linearised equations is called the spectral quasi-linearisation method (SQLM). The SQLM assumes that there is little variation in the solution's approximation at the current iteration level and the one from the prior iteration. Also, the derivatives at the next iteration levels are considered to differ just slightly.

MATHEMATICAL FORMULATIONS

We consider a stretching sheet as the path of a continuous non-linear, two-dimensional, and convective flow of thixotropic nanofluid. The stretched surface's flow is contained inside the domain $y \geq 0$. The stretching velocity of the sheet is $u_w(x) = cx$ (c being a non-negative constant). The amplitude of a constant magnetic field B_0 is parallel to the y −axis (perpendicular to the stretched surface) (shown in Fig. **1**).

The induced magnetic field has been disregarded since the Reynolds number is thought to be negligible. When thixotropic fluid flow equations are subjected to standard boundary layer approximations, they provide:

$$\frac{\partial u}{\partial x}+\frac{\partial v}{\partial y}=0, \tag{1}$$

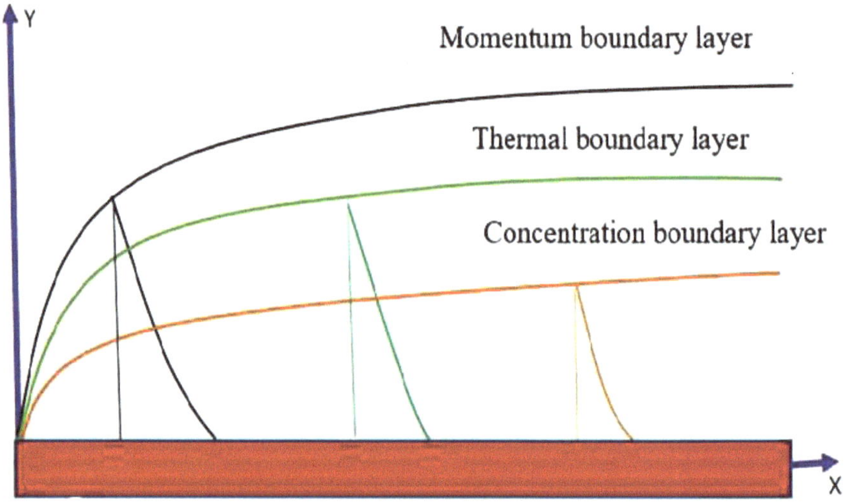

Fig. (1). Physical diagram of the problem under consideration.

$$u\frac{\partial u}{\partial x}+v\frac{\partial u}{\partial y}=\nu\frac{\partial^2 u}{\partial y^2}-\frac{6\Lambda_0}{\rho}\left(\frac{\partial u}{\partial y}\right)^2\frac{\partial^2 u}{\partial y^2}+\frac{4\Lambda_1}{\rho}\left[\frac{\partial u}{\partial y}\frac{\partial^2 u}{\partial y^2}\left(u\frac{\partial^2 u}{\partial x\partial y}+v\frac{\partial^2 u}{\partial y^2}\right)+\left(\frac{\partial u}{\partial y}\right)^2\left(u\frac{\partial^3 u}{\partial x\partial y^2}+v\frac{\partial^3 u}{\partial y^3}+\frac{\partial u}{\partial y}\frac{\partial^2 u}{\partial x\partial y}+\frac{\partial v}{\partial y}\frac{\partial^2 u}{\partial y^2}\right)\right]-\frac{\sigma {B_0}^2}{\rho_f}u+g\beta_1(T-T_\infty)+\beta_2(T-T_\infty)^2+g\beta_3(C-C_\infty)+\beta_4(C-C_\infty)^2 \tag{2}$$

$$u\frac{\partial T}{\partial x}+v\frac{\partial T}{\partial y}=\frac{k_f}{(\rho c_p)_f}\left(\frac{\partial^2 T}{\partial y^2}\right)+\tau D_B\left(\frac{\partial T}{\partial y}\frac{\partial C}{\partial y}\right)+\frac{\tau D_T}{T_\infty}\left(\frac{\partial T}{\partial y}\right)^2-\frac{1}{(\rho c_p)_f}\frac{\partial q_r}{\partial y}, \tag{3}$$

$$u\frac{\partial C}{\partial x}+v\frac{\partial C}{\partial y}=D_B\left(\frac{\partial^2 C}{\partial y^2}\right)+\frac{D_T}{T_\infty}\left(\frac{\partial^2 T}{\partial y^2}\right), \tag{4}$$

The boundary conditions for the flow field are:

$$u=u_w(x)=cx, v=0, T=Tw, C=Cw, at\ y=0$$

$$u\to 0, T\to T_\infty, C\to C_\infty\ as\ y\to\infty. \tag{5}$$

In the aforesaid equations, the x direction velocity component is denoted by u and similarly, the velocity component for y direction is denoted by $v. \nu=(\mu/\rho)_f$ for the kinematic viscosity, Λ_0 and Λ_1 for the material constants, σ for electrical conductivity, β_1 being the linear thermal expansion coefficient and β_2 is nonlinear thermal expansion coefficient, again β_3 is the linear concentration expansion coefficient and β_4 is for nonlinear concentration expansion coefficient, g for gravitational acceleration, ρ_f denotes fluid density, $(c_p)_f$ denotes fluid-specific heat, k_f for thermal conductivity, τ for capacity ratio, D_B and D_T for Brownian and thermophoretic diffusion coefficients, q_r for radiative heat flux, T and C for fluid temperature and fluid concentration, T_∞ is denoted as fluid temperature and C_∞ is for nanoparticle concentration, h_t and h_c are indications of heat transfer and mass diffusion coefficient, respectively.

Here Rosseland's relationship gives:

$$q_r = -\frac{4\sigma^*}{3k^*}\frac{\partial T^4}{\partial y}, \tag{6}$$

where σ^* and $k^*$is the constant of Stefan-Boltzmann and coefficient of mean absorption respectively. As done by Raptis, inside the flow, the temperature difference is seen as being unimportant such that T^4 may represent as a function of temperature in the first degree. This is done by a Taylor series expansion of T^4 about the free stream temperature T_∞, while ignoring higher terms, we obtain

$$T^4 \approx 4T_\infty^3 T - 3T_\infty^4. \tag{7}$$

Thus using Eq.(7) in Eq.(6), we get:

$$q_r = -\frac{16\sigma^* T_\infty^3}{3k^*}\frac{\partial T}{\partial y}. \tag{8}$$

Now Eqs.(3) and (8) gives:

$$u\frac{\partial T}{\partial x} + v\frac{\partial T}{\partial y} = \frac{k_f}{(\rho c_p)_f}\left(\frac{\partial^2 T}{\partial y^2}\right) + \tau D_B\left(\frac{\partial T}{\partial y}\frac{\partial C}{\partial y}\right) + \frac{\tau D_T}{T_\infty}\left(\frac{\partial T}{\partial y}\right)^2 + \frac{1}{(\rho c_p)_f}\frac{16\sigma^* T_\infty^3}{3k^*}\frac{\partial^2 T}{\partial y^2}. \tag{9}$$

Taking into consideration:

$$\eta = \sqrt{\frac{c}{\nu}}y, \theta(\eta) = \frac{T - T_\infty}{T_w - T_\infty}, \phi(\eta) = \frac{C - C_\infty}{C_w - C_\infty},$$

$$u = cxf'(\eta), v = -\sqrt{c\nu}f(\eta), \psi(\eta) = \sqrt{c\nu}xf(\eta), \tag{10}$$

Eq.(1) is merely satisfied and other equations give:

$$f''' - f'^2 + ff'' + \alpha(f'')^2 f''' + \beta(f'f''^2 f''' + f''^4 - ff''f'''^2 - ff''^2 f^{iv}) - Ha^2 f' + \lambda(1 + \beta_t\theta)\theta + \lambda N^*(1 + \beta_c\phi)\phi = 0, \quad \textbf{(11)}$$

$$(1 + Nr)\theta'' + Pr(f\theta' - 2f'\theta) + N_b\theta'\phi' + N_t\theta'^2 = 0, \quad \textbf{(12)}$$

$$\phi'' + Le(f\phi' - 2f'\phi) + \frac{N_t}{N_b}\theta'' = 0, \quad \textbf{(13)}$$

The resultant boundary conditions are:

$$f'(\eta) = 1, f(\eta) = 0, \theta(\eta) = 1, \phi(\eta) = 1 \; at \; \eta = 0$$

$$f'(\eta) = 0, \theta(\eta) = 0, \phi(\eta) = 0 \; at \; \eta \to \infty. \quad \textbf{(14)}$$

Where α and β are the thixotropic parameters, Ha is the Hartmann number, λ is the mixed convection parameter, β_t is the nonlinear convection parameter for temperature and β_c is the nonlinear convection parameter for concentration, N^* is the ratio of concentration to thermal buoyancy forces, Pr is the Prandtl number, Nr is the thermal radiation parameter, N_b is Brownian motion parameter, N_t is thermophoresis variable, Le for Lewis number.

These parameters are outlined as:

$$\alpha = -\frac{6\Lambda_0 c^3 x^2}{\rho v^2}, \beta = \frac{4\Lambda_1 c^4 x^2}{\rho v^2}, Ha = \sqrt{\frac{\sigma}{\rho c}} B_0, \lambda = \frac{Gr}{Re_x^2}, \beta_t = \frac{\beta_2 T_\infty}{\beta_1}$$

$$N^* = \frac{Gr^*}{Gr} = \frac{\beta_3 C_\infty}{\beta_1 T_\infty}, \beta_c = \frac{\beta_4 C_\infty}{\beta_1}, Nr = \frac{16\sigma^* T_\infty^3}{3k_f k^*}, Pr = \frac{v(\rho c_p)_f}{k_f},$$

$$N_b = \frac{\tau D_B}{v}(C_w - C_\infty), N_t = \frac{\tau D_T}{v T_\infty}(T_w - T_\infty), Le = \frac{v}{D_B}, \quad \textbf{(15)}$$

RESULTS AND DISCUSSIONS

Using these similarity transformations, the set of all PDEs has been transformed into a set of ODEs. Using the spectra quasilinearization method (SQLM) concerning these conditions we numerically solved the transformed equation and the effects of velocity, temperature, and nanoparticle concentration are solved numerically. Here we use the values of parameters $\alpha = 0.25$, $\beta = 0.5$, $\lambda = 0.25$, $Ha = 0.5$, $\beta_t = 0.25$, $\beta_c = 0.5$, $N^* = 0.3$, $Pr = 6.8$, $Nt = 0.4$, $Nb = 0.3$, $Le = 0.2$, $Nr = 0.5$. The present results using SQLM with previously published works are in excellent agreement with Khan and Pop [2] (see Table **1**).

Table 1. Verification of the present results using SQLM with Khan and Pop [14] of Nusselt number.

Nb	Nt	Pr	Le	Khan and Pop[14]		Present Result (SQLM)	
				$-\theta'(0)$	$-\phi'(0)$	$-\theta'(0)$	$-\phi'(0)$
0.1	0.1	10	10	0.9524	2.1294	0.952455	2.129345
0.2	0.2	10	10	0.3654	2.5152	0.365398	2.515119
0.3	0.3	10	10	0.1355	2.6088	0.135566	2.608782
0.4	0.4	10	10	0.0495	2.6038	0.049578	2.603786
0.5	0.5	10	10	0.0179	2.5731	0.017945	2.573302

Figures are prepared on velocity fields for physical deviation of parameters like magnetic parameter Ha, α, and β as the thixotropic parameters and mixed convection λ. Figs. (**2a** and **2b**) show the variations of α and β as thixotropic parameters for the velocity profile by keeping the other variables constant. The velocity increases when α and β increase. The layer thickness is increased for higher values of α and β. In Fig. (**3a**), it is shown that the velocity decreases if Ha increases. The layer thickness decreases if Ha increases. For increasing values of Ha, Lorentz force increases which makes the fluid more resistant to motion for which the velocity decreases. Fig. (**3b**) shows the velocity increases if λ increases. The layer thickness increases if λ increases.

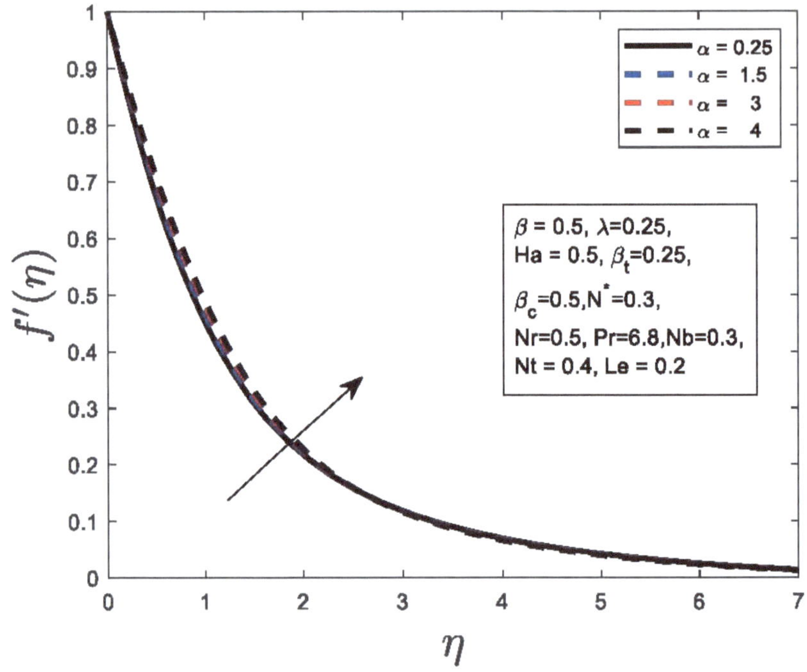

Fig. (2a). Effect of α in velocity profile.

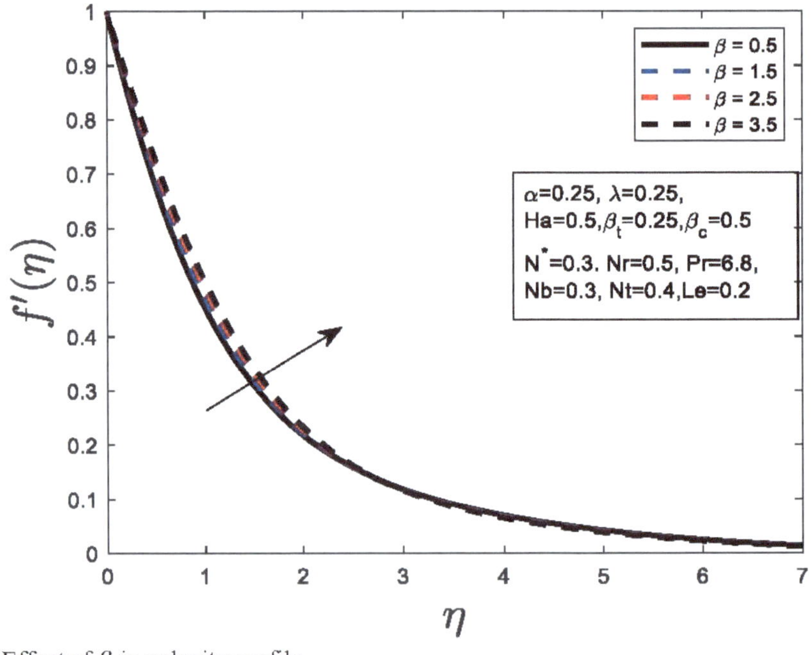

Fig. (2b). Effect of β in velocity profile.

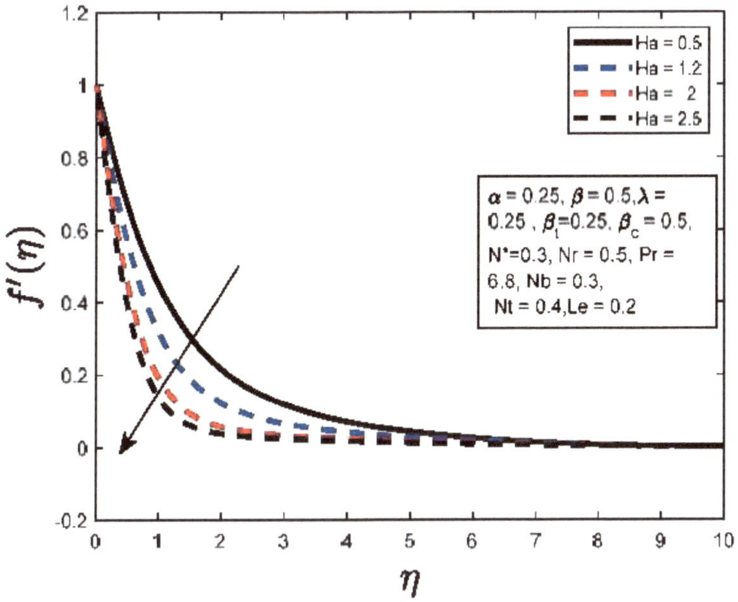

Fig. (3a). Effect of Ha in the velocity profile.

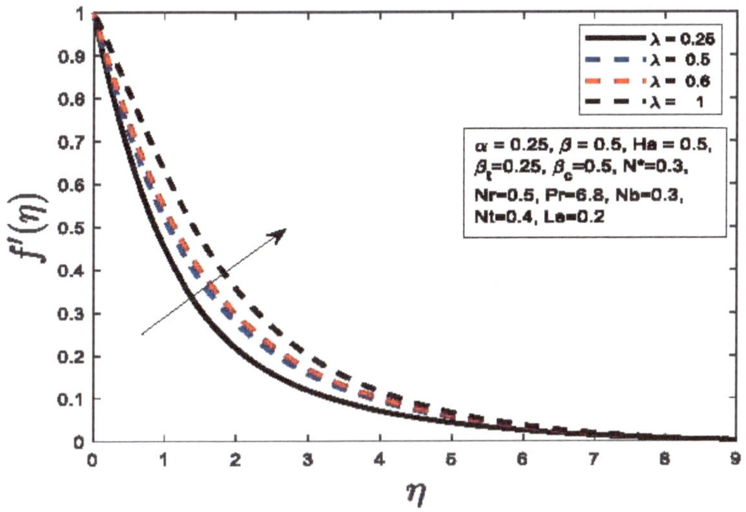

Fig. (3b). Effect of λ in velocity profile.

Figures are prepared for physical deviation of parameters like Hartmann number Ha, Pr as the Prandtl number, Nt as a Thermophoresis parameter, Nb as the Brownian motion parameter on temperature distributions. In Fig. (**4a**), it is noted that there is an increase in temperature if Ha increases. Also, an increase in the thickness of the boundary layer is observed if Ha increases. Fig. (**4b**) shows that temperature decreases if Pr increases. The thickness of the boundary layer decreases if Pr increases. Since thermal diffusivity is correlated to the momentum by Pr, if Pr increases it causes a decrease in thermal diffusivity, so temperature decreases.

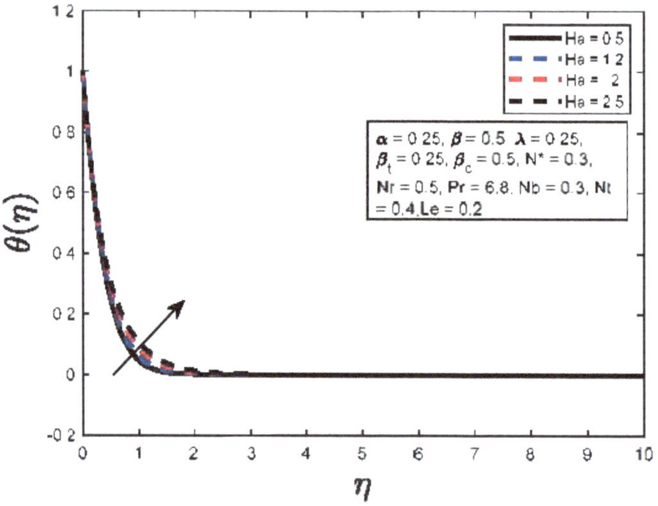

Fig. (4a). Effect of Ha on temperature profile.

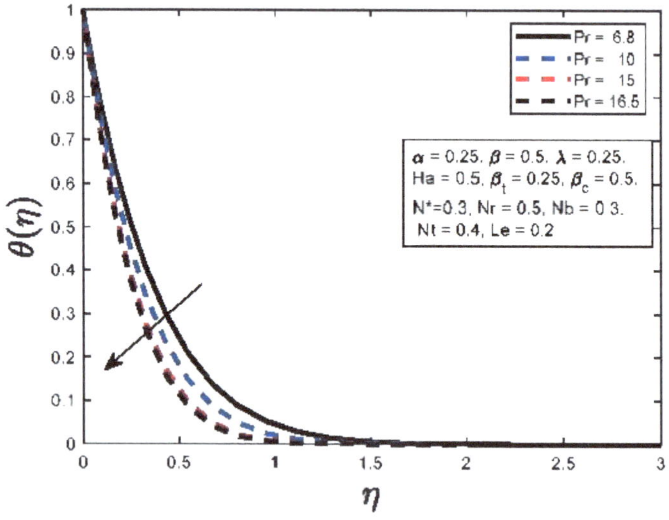

Fig. (4b). Effect of Pr on the temperature profile.

Fig. (**5a**) shows that by increasing the value of Nt, temperature increases. Also, the width of the boundary layer increases if Nt increases. The thermophoresis phenomenon deals with the transmission of heated particles from hot to cold surfaces which increases the fluid temperature. Fig. (**5b**) shows temperature increases if Nb increases. Conclusions drawn for Nb parameter are that the thickness of the boundary layer increases if Nb increases. If Nb increases then the Brownian motion of the particles in fluid is increased, so the random motion of the particles increases, so temperature increases.

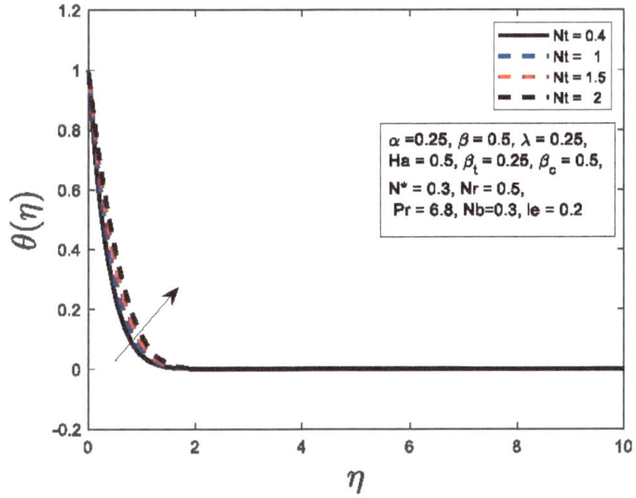

Fig. (5a). Effect of Nt in temperature profile.

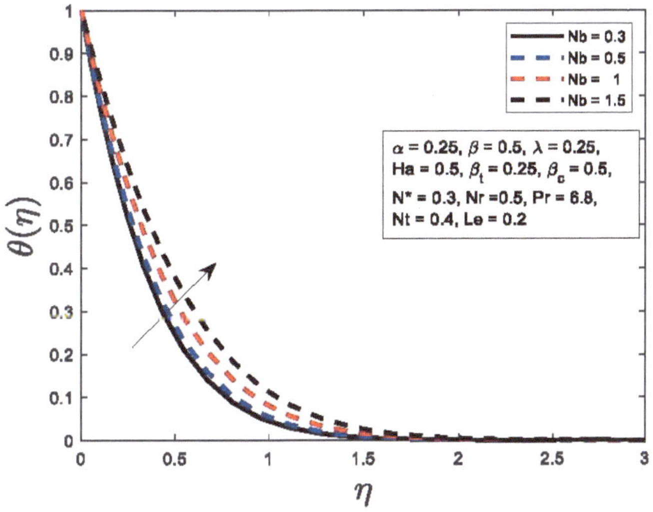

Fig. (5b). Effect of Nb in temperature profile.

Figures are prepared for the physical deviation of parameters like Brownian motion which is denoted by Nb, Lewis number as Le, and Thermophoresis parameter as Nt on temperature distributions. Fig. **(6a)** depicts that the concentration field diminishes when Nb increases. Also, the thickness of the boundary layer diminishes when Nb increases. If Nb increases then the collision between the fluid particles increases, so the concentration decreases. From Fig. **(6b)** it is clear for Nt that the concentration enhances when Nt increases. The conclusion drawn for Nt parameter in the concentration profile is that the thickness of the boundary layer enhances when Nt increases.

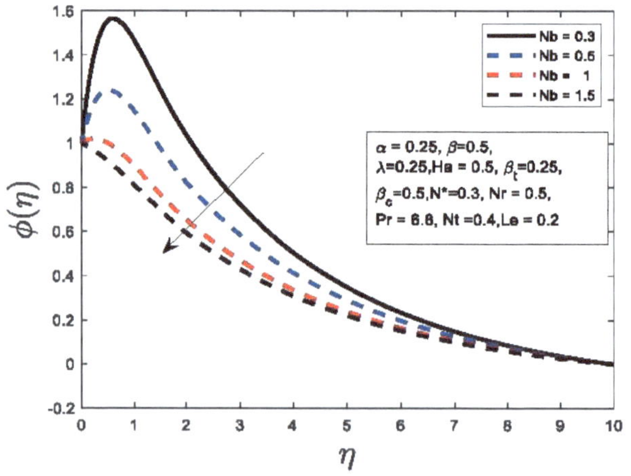

Fig. (6a). Effect of Nb in concentration profile.

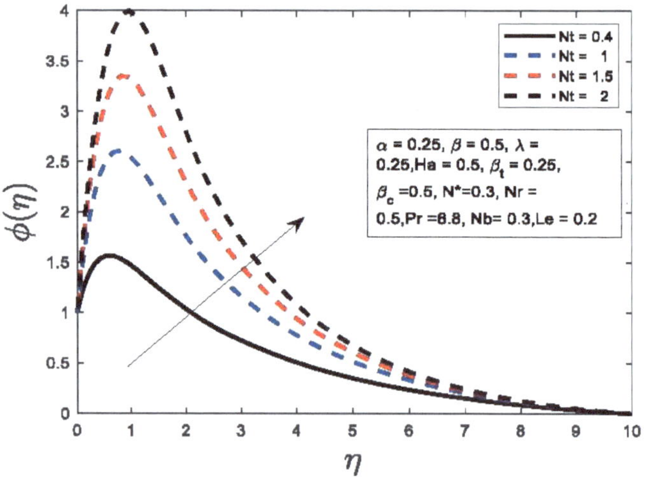

Fig. (6b). Effect of Nt in concentration profile.

From Fig. (**7a**), it has been observed for Nr that if Nr increases, there is an increase in temperature. Also, we note that the width of the boundary layer increases if Nr increases. As Nr is thermal radiation parameter, if Nr increases then thermal radiation increases. So the temperature increases. Next from Fig. (**7b**) it is noted that concentration decreases when Le increases. The width of the boundary layer decreases while Le increases. Lewis number is the ratio of Schmidt number to Prandtl number (*i.e.*, the ratio of thermal diffusivity to mass diffusivity). So when $Le = 1.0$, then at the same rate the heat and species will diffuse. When Le increases, the heat will diffuse faster than the species. So, the concentration and boundary layer thickness decrease.

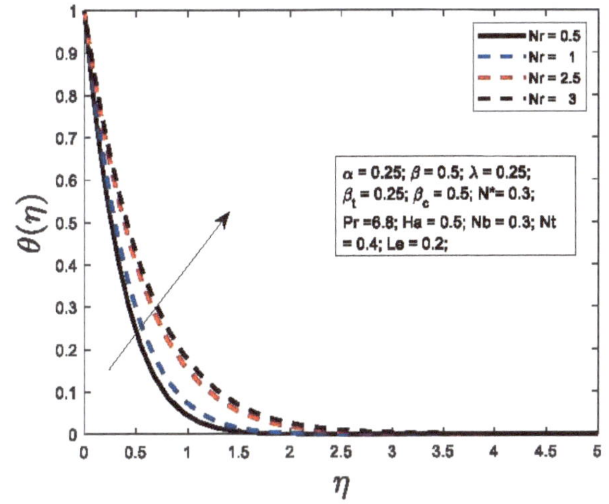

Fig. (7a). Effect of Nr on temperature profile.

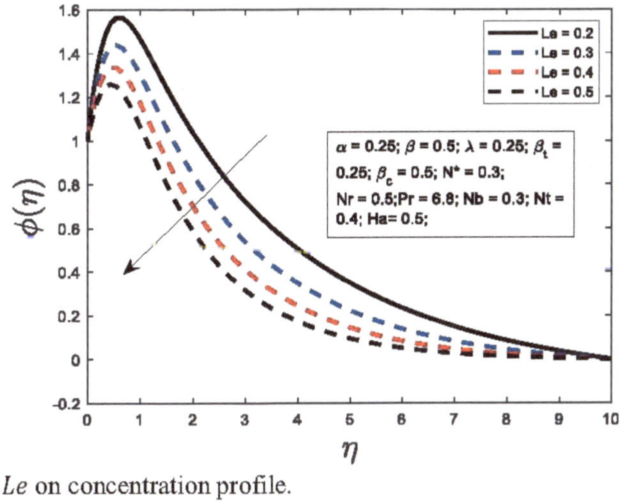

Fig. (7b). Effect of Le on concentration profile.

CONCLUSION

The analysis of thixotropic nanofluid's MHD convective flow under Newtonian heat and mass conditions is investigated. The main outcomes of this research are as follows:

- Velocity increases for higher values of thixotropic parameters (α and β)and mixed convection flow parameter (λ).
- For higher values of Hartmann number (Ha), the temperature increases but velocity decreases.
- If the Prandtl number (Pr) increases, the temperature decreases.
- Concentration decreases for higher values of Nb and increases for higher values of Nt.

REFERENCES

[1] C. Fetecau, J. Zierep, R. Bohning, and C. Fetecau, "On the energetic balance for the flow of an Oldroyd-B fluid due to a flat plate subject to a time dependent shear stress", *Computers & Mathematics with Applications.*, vol. 60, no. 1, pp. 74-82, 2010.
http://dx.doi.org/10.1016/j.camwa.2010.04.031

[2] S. Wang, and W. Tan, "Stability analysis of soret-driven double-diffusive convection of Maxwell fluid in a porous medium", *Int. J. Heat Fluid Flow,* vol. 32, no. 1, pp. 88-94, 2011.
http://dx.doi.org/10.1016/j.ijheatfluidflow.2010.10.005

[3] S. Qayyum, T. Hayat, A. Alsaedi, and B. Ahmad, "MHD nonlinear convective flow of thixotropic nanofluid with chemical reaction and Newtonian heat and mass conditions", *Results Phys.*, vol. 7, pp. 2124-2133, 2017.
http://dx.doi.org/10.1016/j.rinp.2017.06.010

[4] O.K. Koriko, N.A. Shah, S. Saleem, J.D. Chung, A.J. Omowaye, and T. Oreyeni, "Exploration of bioconvection flow of MHD thixotropic nanofluid past a vertical surface coexisting with both nanoparticles and gyrotactic microorganisms", *Sci. Rep.,* vol. 11, no. 1, p. 16627, 2021.
http://dx.doi.org/10.1038/s41598-021-96185-y PMID: 34404877

[5] R. Mehta, V.S. Chouhan, and T. Mehta, "Mhd flow of nanofluids in the presence of porous media, radiation and heat generation through a vertical channel", *J. Phys. Conf. Ser.,* vol. 1504, no. 1, p. 012008, 2020.
http://dx.doi.org/10.1088/1742-6596/1504/1/012008

[6] G. Dharmaiah, S. Dinarvand, J.L. Rama Prasad, S. Noeiaghdam, and M. Abdollahzadeh, "Non-homogeneous two-component buongiorno model for nanofluid flow toward Howarth's wavy cylinder with activation energy", *Results in Engineering,* vol. 17, no. Jan, p. 100879, 2023.
http://dx.doi.org/10.1016/j.rineng.2023.100879

[7] A. Shafiq, A.B. Colak, and T.N. Sandhu, "Significance of bioconvective flow of MHD thixotropic nanofluid passing through a vertical surface by machine learning algorithm", *Chin. J. Phys.,* vol. 80, pp. 427-444, 2022.

http://dx.doi.org/10.1016/j.cjph.2022.08.008

[8] M. Waqas, A.S. Dogonchi, S.A. Shehzad, M.I. Khan, T. Hayat, and A. Alsaedi, "Nonlinear convection and joule heating impacts in magneto-thixotropic nanofluid stratified flow by convectively heated variable thicked surface", *J. Mol. Liq.,* vol. 300, p. 111945, 2020.

http://dx.doi.org/10.1016/j.molliq.2019.111945

[9] K.A. Kumar, V. Sugunamma, N. Sandeep, and J.V.R. Reddy, "Numerical examination of MHD nonlinear radiative slip motion of non-Newtonian fluid across a stretching sheet in the presence of a porous medium", *Heat Transf. Res.,* vol. 50, no. 12, pp. 1163-1181, 2019.

http://dx.doi.org/10.1615/HeatTransRes.2018026700

[10] M. Bibi, A. Zeeshan, and M.Y. Malik, "Numerical analysis of unsteady flow of three-dimensional Williamson fluid-particle suspension with MHD and nonlinear thermal radiations", *Eur. Phys. J. Plus,* vol. 135, no. 10, p. 850, 2020.

http://dx.doi.org/10.1140/epjp/s13360-020-00857-z

[11] H. Kataria, A.S. Mittal, and M. Mistry, "Effect of nonlinear radiation on entropy optimised MHD fluid flow", *Int. J. Ambient Energy,* vol. 43, no. 1, pp. 6909-6918, 2022.

http://dx.doi.org/10.1080/01430750.2022.2059000

[12] K. Das, "Effect of chemical reaction and thermal radiation on heat and mass transfer flow of MHD micropolar fluid in a rotating frame of reference", *Int. J. Heat Mass Transf.,* vol. 54, no. 15-16, pp. 3505-3513, 2011.

http://dx.doi.org/10.1016/j.ijheatmasstransfer.2011.03.035

[13] B. Jamil, M.S. Anwar, A. Rasheed, and M. Irfan, "MHD Maxwell flow modeled by fractional derivatives with chemical reaction and thermal radiation", *Zhongguo Wuli Xuekan,* vol. 67, pp. 512-533, 2020.

http://dx.doi.org/10.1016/j.cjph.2020.08.012

[14] W.A. Khan, and I. Pop, " Boundary-layer flow of a nanofluid past a stretching sheet", Int *J Heat Mass Transf,* vol, 53, 2477 2483, 2010.

CHAPTER 8

Effect of MHD Viscous Nanofluid Flow in the Presence of Internal Heat Generation

Nibedita Mandal[1], Sewli Chatterjee[2] and Hiranmoy Mondal[1,*]

[1] *Department of Applied Mathematics, Maulana Abul Kalam Azad University of Technology, Haringhata, Nadia, West Bengal, India*

[2] *Department of Mathematics, Turku Hansda Lapsa Hemram Mahavidyalay, (Under Burdwan University), Mallarpur, West Bengal, India*

Abstract: In this current study, we investigated the boundary layer steady nanofluid flow which is viscous. Utilizing the spectral quasi-linearization method (SQLM), the mathematical model has been solved. To analyze the convergence of the numerical method, we calculated the residual errors that are approximately less by [10^{-8}]. The local Sherwood number, the coefficient of the rate of heat transfer, and the local drag force coefficient along with various flow parameters are analyzed numerically and graphically. The influence of thermophoresis and Brownian motion has also been discussed briefly.

Keywords: Brownian motion, Heat generation, Nanofluids, Thermophoresis.

INTRODUCTION

Nanofluids are such a fluid in which nano-sized particles are suspended in a base fluid to such an extent that no sedimentation can take place. Nanofluid is an adhesive solution of nano-molecule in the base fluid. All kinds of metal, carbon nanotubes, carbides, oxides, *etc.* are used as nanoparticles, and water, oil, ethylene glycol, and others are used as the base fluid. Nanofluids are known for their unprecedented prominences which are inequivalent from the base fluid and involve thermophysical characteristics like density, specific heat, thermal conductivity, viscosity, *etc.* The thermophysical features of fluid are changed after adding nanoparticles to it. Thence nanofluids are used for their improved thermal features as cooling in heat conduct tackle like heat exchangers, electronic coolant systems, and convectors. A magnetohydrodynamic field produces currents in a moving conducting fluid flow that produces energy and generates a magnetic field effect. Most industrial procedure entangles the stream of reactive liquids. Considering the

[*]**Corresponding author Hiranmoy Mondal:** Department of Applied Mathematics, Maulana Abul Kalam Azad University of Technology, Haringhata, Nadia, West Bengal; E-mail: hiranmoymondal@yahoo.co.in

Sabyasachi Mondal (Ed.)

nanofluids boundary layer stream which is incompressible and viscous in nature, we have investigated heat and mass transfer throughout the stretching lamina. Such streams are confronted in industrial procedures like polymer extrusion, plastic, metal ejection, and glass blowing studied by Makinde and Sibanda [1], Hayat *et al.* [2], and Sohut *et al.* [3].

The rate of cooling and the chemical reactions in manufacturing industries measure the grade of products studied by Yongjae [4] and Boekel [5]. In a chemical reaction in the food product industry in a significant area, heat can be produced/absorbed by a chemical transformation examined by Olanrewaju *et al.* [6], so controlling heat transportation in a system is very important.

Bhattacharyya [7] applied an effective magnetic field to control the heat transportation rate. Nanofluid is considerably efficient in transporting heat and mass through the fluid flow. It is very useful in the industrial area and medical sector to apply Nanofluid instead of ordinary fluid explained by Abu-Nada [8] and Mohammed *et al.* [9]. Nanofluids are the compositions of nano-sized particles and base fluids, typically metal or metal-oxide, carbides, SWCNTs, or single-wall carbon nanotubes, and MWCNTs or multi-wall carbon nanotubes are mixed in base fluids like water, oil, ethylene glycol, *etc.* Application of nanofluids occupies an extensive area in solar water-heating, cooling techniques, electronics equipment, and the enhancement of the efficiency of diesel-based generators examined by Hussein and Hussain [10], Hayat *et al.* [11] and Mahanthesh *et al.* [12].

The issue of the production of heat in the system has been exhibited to be very concerning, so giving attention to these physical phenomena of heat production is considerable. The fluid velocity attains the extreme value in case of no thermal radiation. Palani *et al.* [13] scrutinized the same type of investigation excluding energy transportation, we detect that a simultaneous study of velocity profile along with energy and mass transportation may be much more effective as they are correlated to each other.

Mlamuli *et al.* [14] found that the sequel of the Brownian motion parameter and the thermophoresis parameter have a revoking reaction on the velocity profile and concentration profile, whereas both raise the fluid temperature. Sarkar *et al.* [15] examined that the momentum boundary layer reduces as the magnetic parameter increases.

Mondal *et al.* [16] investigated the power-law fluid and solved the numerical model by using the spectral quasi-linearisation method. The impact of thermal radiation on nano-fluid flow with higher-order chemical reactions was studied by Sarkar *et al.* [17].

The current study aims to investigate the momentum, energy, and mass transportation of a nanofluid. Important factors like internal heat production, the effect of the magnetic field, dissipation, buoyancy force, and thermophoretic effect have been taken into account to identify their impact on the fluid velocity profile, thermal energy, and mass transportation characteristics. Convective boundary conditions are considered in this present study. Here spectral quasi-linearization method is used to solve the model equations.

FORMULATION OF THE PROBLEM

Along the way of the mainstream of the lamina, $x - axis$ has been considered. The direction of $y - axis$ is perpendicular to the lamina with velocity components u, v along $x - axis$ and $y - axis$, respectively see (Fig. 1).

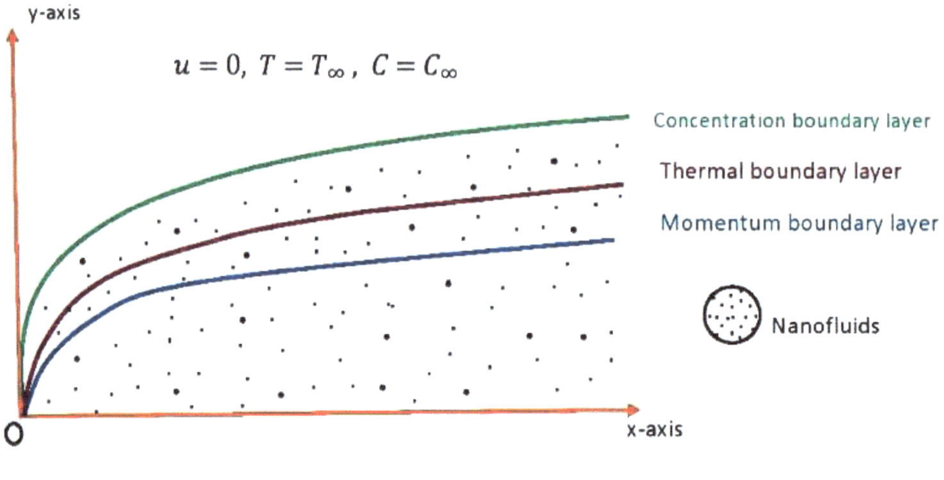

Fig. (1). Schematic diagram.

The ruling equations of a two-dimensional stream considering the boundary layer approximation are presented in this fashion:

$$\frac{\partial u}{\partial x}+\frac{\partial v}{\partial y}=0 \tag{1}$$

$$u\frac{\partial u}{\partial x}+v\frac{\partial u}{\partial y}=\nu\frac{\partial^2 u}{\partial y^2}-\frac{\sigma B^2(x)}{\rho}u+g\beta_T(T-T_\infty)+g\beta_C(C-C_\infty) \tag{2}$$

$$u\frac{\partial T}{\partial x}+v\frac{\partial T}{\partial y}=\alpha\frac{\partial^2 T}{\partial y^2}-\frac{\nu}{C_p}\left(\frac{\partial u}{\partial y}\right)^2+\tau\left[D_B\frac{\partial T}{\partial y}\frac{\partial C}{\partial y}+\frac{D_T}{T_\infty}\left(\frac{\partial T}{\partial y}\right)^2\right]+\frac{Q_0}{\rho c_p}(T-T_\infty) \tag{3}$$

$$u\frac{\partial C}{\partial x}+v\frac{\partial C}{\partial y}=D_B\frac{\partial^2 C}{\partial y^2}+\frac{D_T}{T_\infty}\frac{\partial^2 T}{\partial y^2} \tag{4}$$

The relevant boundary conditions are:

$$u=U_0,\ \ v=0\text{, when } y=0,\ t>0$$

$$u\to 0,\ T\to T_\infty, C\to C_\infty\ when\ y\to\infty,\ t>0$$

$$D_B\frac{\partial C}{\partial y}+\frac{D_T}{T_\infty}\left(\frac{\partial T}{\partial y}\right)=0,\ \ at\ y=0,\ t>0 \text{ and } T\to T_\infty\ when\ \ y\to\infty,\ t>0.$$

ANALYTICAL SOLUTION

Stream function $\Psi(x,y)$ worthwhile for the continuity equation:

$$u=\frac{\partial\Psi}{\partial x}, v=-\frac{\partial\Psi}{\partial y}. \tag{5}$$

Considering the similarity conversion:

$$\eta=y\sqrt{\frac{\lambda}{\nu}},\quad \Psi(x,y)=x\sqrt{\nu\lambda}f(\eta)$$

Substituting all the above values, Eqs. number (2), (3), (4) transformed as follows,

$$f'''+\frac{1}{2}f''f-Mf'+\lambda_1\theta+\lambda_2\emptyset=0 \tag{6}$$

$$\theta''+P_r\left[\frac{1}{2}f\theta'+E_cf''^2+Nb\theta'\emptyset'+Nt\theta'^2+\delta\theta\right]=0 \tag{7}$$

$$\emptyset''+\frac{1}{2}Scf\emptyset'+\frac{Nt}{Nb}\theta''=0 \tag{8}$$

$$\left.\begin{aligned} f'(0) = 1,\ f(0) = 0,\ f'(\infty) \to 0 \\ \theta'(0) = -Bi\big(1 - \theta(0)\big), \theta(\infty) \to 0 \\ Nb\emptyset'(0) + Nt\theta'(0) = 0, \emptyset(\infty) \to 0 \end{aligned}\right\} \quad \textbf{(9)}$$

NUMERICAL METHODS (SPECTRAL QUASI-LINEARIZATION METHODS)

Amalgamated with boundary condition (9), the nonlinear ordinary differential Eqs. (6), (7), and (8) were solved numerically by applying the spectral quasilinearization method (SQLM).

$$F \equiv f''' + \frac{1}{2}f''f - Mf' + \lambda_1\theta + \lambda_2\emptyset \quad \textbf{(10)}$$

$$\bar{\theta} \equiv \theta'' + Pr\left[\frac{1}{2}f\theta' + Ecf''^2 + Nb\theta'\emptyset' + Nt\theta'^2 + \delta\theta\right] \quad \textbf{(11)}$$

$$\bar{\emptyset} = \emptyset'' + \frac{1}{2}Scf\theta' + \frac{Nt}{Nb}\theta'' \quad \textbf{(12)}$$

We establish the error from the iterative method for the Eqs. (6)-(8) as given below:

$$a_{0,\ r}f'''_{r+1} + a_{1,\ r}f''_{r+1} + a_{2,\ r}f'_{r+1} + a_{3,\ r}f_{r+1} + a_{4,\ r}\theta_{r+1} + a_{5,\ r}\emptyset_{r+1} = R_F,$$

$$b_{0,\ r}\theta''_{r+1} + b_{1,\ r}\theta'_{r+1} + b_{2,\ r}\theta_{r+1} + b_{3,\ r}f''_{r+1} + b_{4,\ r}f_{r+1} + b_{5,\ r}\emptyset'_{r+1} = R_{\bar{\theta}},$$

$$c_{0,\ r}\emptyset''_{r+1} + c_{1,\ r}\emptyset'_{r+1} + c_{2,\ r}f_{r+1} + c_{3,\ r}\theta''_{r+1} = R_{\bar{\emptyset}}$$

With suitable boundary conditions:

$$f_{r+1}(0) = 0,\ \ f'_{r+1}(0) = 1,\ \ f'_{r+1}(\infty) \to 0,$$

$$\theta'_{r+1}(0) = -Bi(1 - \theta_r(0)),\ \ \theta_{r+1}(\infty) \to 0,$$

$$Nb\emptyset'_{r+1}(0) + Nb\theta'_{r+1}(0) = 0, \qquad \emptyset_{r+1}(\infty) \to 0$$

The coefficients appeared in the above equations are given as:

$$a_{0,\ r} = \frac{\partial F}{\partial f'''} = 1,\ a_{1,\ r} = \frac{\partial F}{\partial f''} = \frac{1}{2}f_r,\ a_{2,\ r} = \frac{\partial F}{\partial f'} = -M,\ a_{3,\ r} = \frac{\partial F}{\partial f} = \frac{1}{2}f_r'',$$
$$a_{4,\ r} = \frac{\partial F}{\partial \theta} = \lambda_1,$$

$$a_{5,\ r} = \frac{\partial F}{\partial \emptyset} = \lambda_2,$$

$$b_{0,\ r} = \frac{\partial \bar{\theta}}{\partial \theta''} = 1,\ b_{1,\ r} = \frac{\partial \bar{\theta}}{\partial \theta'} = Pr\left[\frac{1}{2}f + Nb\emptyset' + 2Nt\theta'\right],\ b_{2,\ r} = \frac{\partial \bar{\theta}}{\partial \theta} = Pr\delta,$$

$$b_{3,\ r} = \frac{\partial \bar{\theta}}{\partial f''} = 2PrEcf'',\ b_{4,\ r} = \frac{\partial \bar{\theta}}{\partial f} = \frac{1}{2}Pr\theta',\ b_{5,\ r} = \frac{\partial \bar{\theta}}{\partial \emptyset'} = PrNb\theta'$$

$$c_{0,\ r} = \frac{\partial \bar{\emptyset}}{\partial \emptyset''} = 1,\ c_{1,\ r} = \frac{\partial \bar{\emptyset}}{\partial \emptyset'} = \frac{1}{2}Scf,\ c_{2,\ r} = \frac{\partial \bar{\emptyset}}{\partial f} = \frac{1}{2}Sc\emptyset',\ c_{3,\ r} = \frac{\partial \bar{\emptyset}}{\partial \theta''} = \frac{Nt}{Nb}$$

The initial guess that satisfies the boundary conditions is defined as follows:

$$f_0(\eta) = 1 - e^{-\eta},\ \theta_0(\eta) = \frac{Bi}{1+Bi}e^{-\eta},\ \emptyset_0(\eta) = -\frac{Nt}{Nb}\frac{Bi}{1+Bi}e^{-\eta}$$

In order to use the Spectral Quasi-Linearization method of nonlinear ordinary differential equations, the domain $0 \leq \eta \leq L_x$ is converted into the computational domain $-1 \leq x \leq 1$ using the metamorphosis, we get,

$$\frac{dF_r^{(1)}}{dn_j}(\eta) = \sum_{k=0}^{n} D1f(\eta_k) = D1F_m, \qquad j = 0, 1, 2, \ldots\ldots, N$$

$$\text{Where } F = [f(\eta_0), f(\eta_1), f(\eta_2), \ldots\ldots f(\eta_N),]^T.$$

By utilizing the scaled differential matrix on (10)-(12), we will get;

$$A_{11}f_{r+1} + A_{12}\theta_{r+1} + A_{13}\emptyset_{r+1} = R_f,$$

$$A_{21}f_{r+1} + A_{22}\theta_{r+1} + A_{23}\emptyset_{r+1} = R_\theta,$$

$$A_{31}f_{r+1} + A_{32}\theta_{r+1} + A_{33}\emptyset_{r+1} = R_\emptyset.$$

That can be written in the matrix form given as follows:

$$\begin{bmatrix} A_{11} & A_{12} & A_{13} \\ A_{21} & A_{22} & A_{23} \\ A_{31} & A_{32} & A_{33} \end{bmatrix} \begin{bmatrix} f_{r+1} \\ \theta_{r+1} \\ \emptyset_{r+1} \end{bmatrix} = \begin{bmatrix} R_f \\ R_\theta \\ R_\emptyset \end{bmatrix},$$

RESULT AND DISCUSSION

In this current article, the impacts of various governing fluid parameters on a hydromagnetic nanofluid flow over a non-linear stretching lamina have been demonstrated. The mathematical model has been solved numerically by using the spectral quasilinearization method, and the outcomes are presented diagrammatically in Figs. (**2–6**). Conclusions are given below to analyze the impacts of different types of physical quantities of interest that have significant effects. The validations of this model using the spectral quasilinearisation methods are presented in Table **1** and are found to be in excellent agreement.

Table 1. Comparison with Patil *et al.* [22] of local Nusselt number for different values of Prandtl number.

Pr	Patil *el al.* [22]	Present Results (SQLM)
0.7	0.349235	0.3492456
1.0	0.443748	0.4437567
2.0	0.683258	0.6832678
7.0	1.387033	1.3870345
10.0	1.680293	1.6802988
100.0	5.544633	5.5446567

BROWNIAN MOTION (*Nb*)

The sequel of Brownian motion on the velocity profiles and concentration profiles are shown in Figs. (**2a** and **2b**). The Brownian motion (Nb) is an arbitrary 'inconclusive' gesticulation of the particles drowned within a liquid caused by the friction between the moving particles contained in the fluid. The momentum boundary layer becomes tenuous in a hike of the Brownian motion. The fluid stuck to a molecule is pulled toward the way of that molecule. At this moment, viscous forces resist the motion of the particle in the liquid examined by Uma *et al.* [18]. All-inclusive, the velocity of the liquid is reduced. The temperature profile increases at the increase of the value of Brownian motion. The solutal boundary layer diminishes with the enhancement of the Brownian motion parameter. The movement of particles is boosted by an upsurge in the Brownian motion parameter (Nb). This causes the heating of the boundary layer and nanoparticles to depart from the surfaces within the immobile fluid. This raises the attestation of the molecule afield to the lamina, causing the depletion of the concentration of the fluid experimented by Goyal *et al.* [19]. The outcomes have been demonstrated in diagram 2(c). Analogous consequences were acquired by Dhlamini *et al.* [20] and Mabood *et al.* [21].

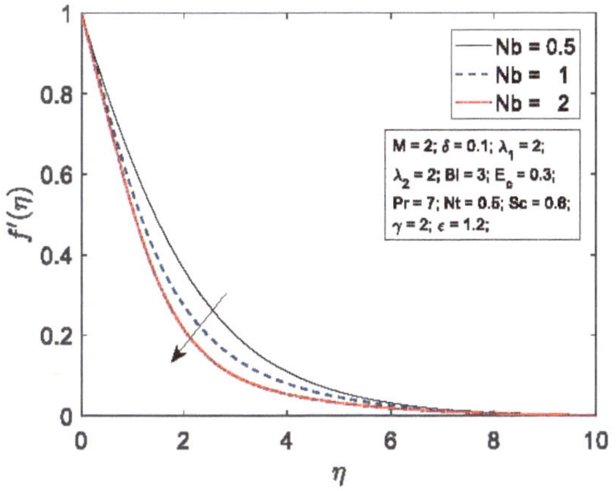

Fig. (2a). Impact of Brownian motion parameter on velocity profile.

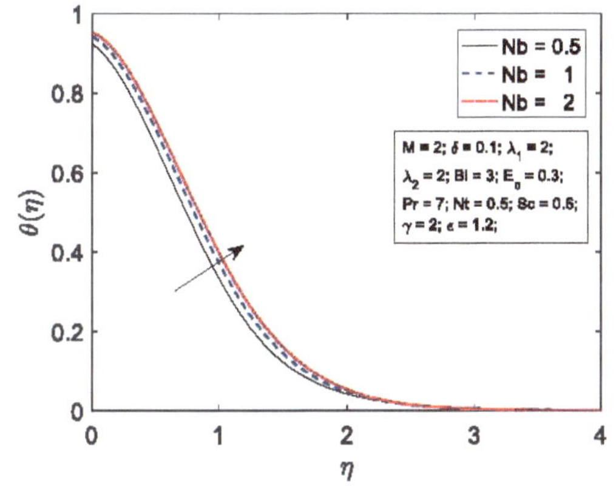

Fig. (2b). Impact of Brownian motion parameter on temperature profile.

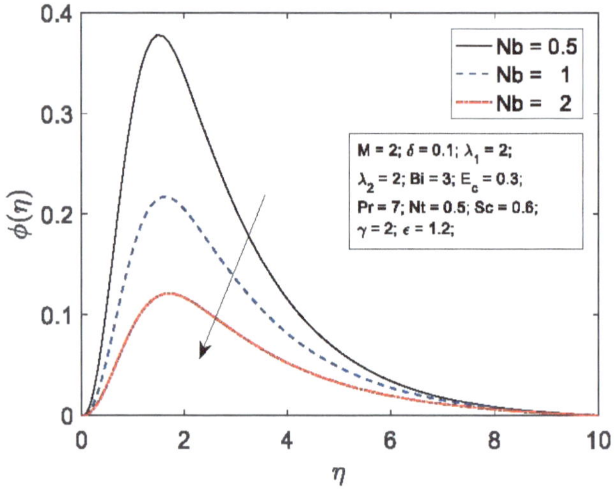

Fig. (2c). Impact of Brownian motion parameter on concentration profile.

THERMOPHORESIS (*Nt*)

Figs. (**3a**, **3b** and **3c**) depict the velocity profiles $f'(\eta)$, temperature profiles $\theta(\eta)$ and concentration profiles $\varphi(\eta)$ for\ different values of thermophoresis parameter *Nt*. It can be seen that a hike in *Nt* causes to rise in both the temperature and nanoparticle concentration. Fluid velocity is also enhanced due to the increment of the thermophoresis parameter.

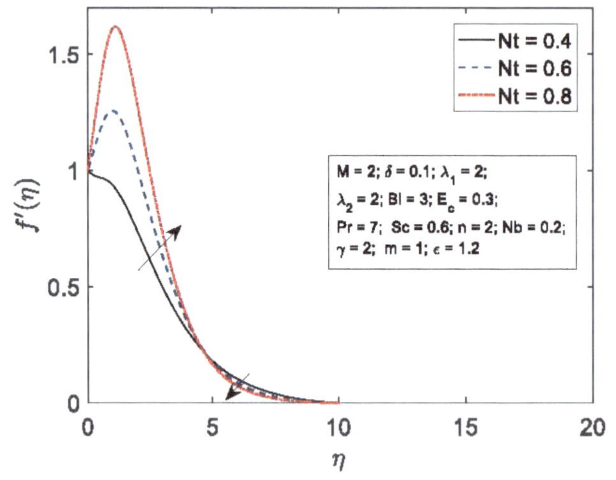

Fig. (3a). Impact of Thermophoresis parameter on velocity profile.

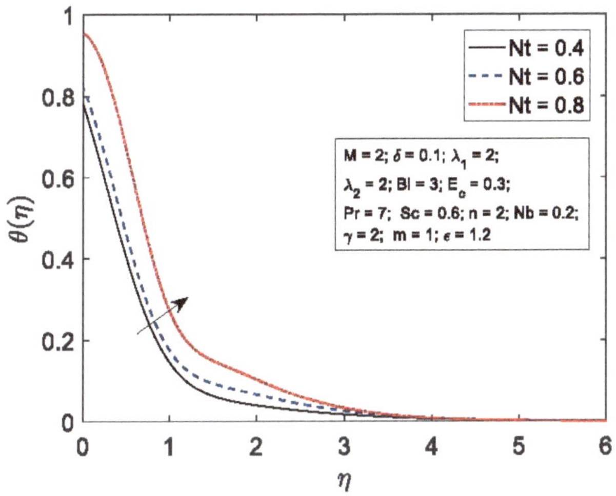

Fig. (3b). Impact of Thermophoresis parameter on temperature profile.

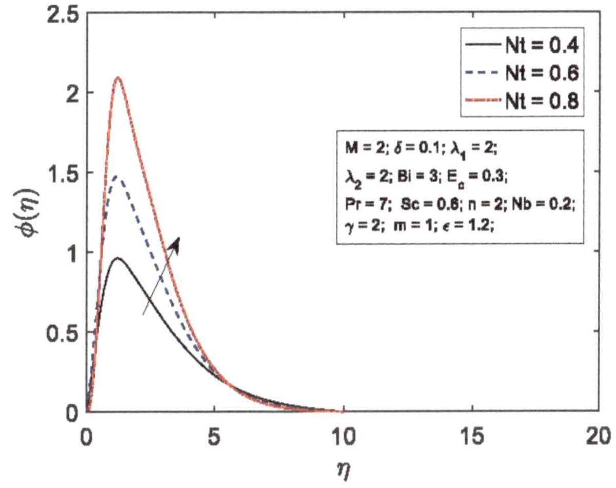

Fig. (3c). Impact of Thermophoresis parameter on concentration layout.

SCHMIDT NUMBER (Sc)

Schmidt number Sc is a non-dimensional number that is defined as the ratio of dynamic viscosity and the mass diffusivity of the fluid. If $Sc > 1$ then the viscosity dominates the thermal diffusivity, on the hand, $Sc < 1$ the thermal diffusivity dominates the viscosity. Figs. (**4a** and **4b**) portray the variation of thermal and concentration layouts, orderly, for different values of Sc. Fig. (**4a**) represents that the temperature increases with the enhancement of the Schmidt number, but opposingly, concentration decreases.

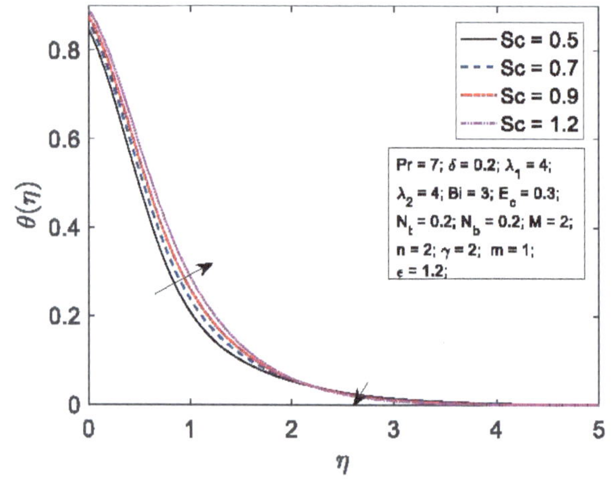

Fig. (4a). Impact of Schmidt Number on temperature profile.

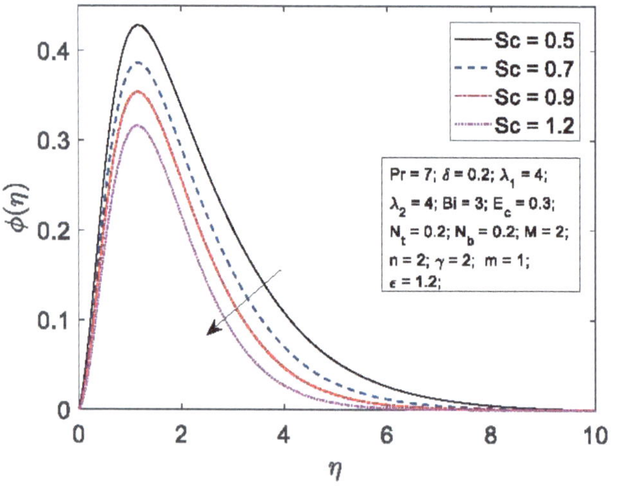

Fig. (4b). Impact of Schmidt Number on concentration layout.

PRANDTL NUMBER (*Pr*)

The Prandtl number (Pr) is non-dimensional, which is defined as the ratio of the dynamic viscosity to the thermal diffusivity of the fluid. The mathematical expression of the Prandtl number is given by *Pr=ν/α=(momentum diffusivity)/(thermal diffusivity)*. The sequel of Prandtl number (Pr) on the heat transfer process is shown by Fig. (**5**). This figure depicts that when the value of the Prandtl number (Pr) increases, the temperature distribution decreases, as the thermal boundary layer broadness diminishes with an upsurge in the Prandtl number.

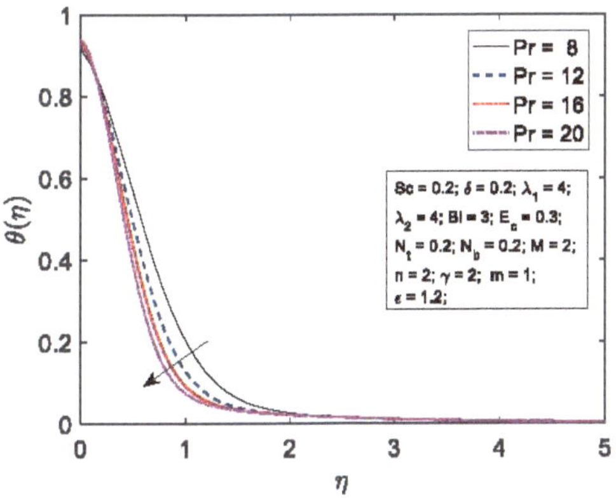

Fig. (5). Sequel of Prandtl number on thermal boundary layout.

MAGNETIC FIELD PARAMETER (*M*)

The effect of magnetic field parameter on the velocity profiles and the temperature profiles have been portrayed in Figs. (**6a** and **6b**). Fig. (**6a**) shows that the velocity diminishes by the increment of the magnetic parameter. Magnetic field generates a repellent force known as Lorentz force which opposes the velocity of the fluid. Due to this phenomenon, the velocity profile is reduced but the temperature is enhanced.

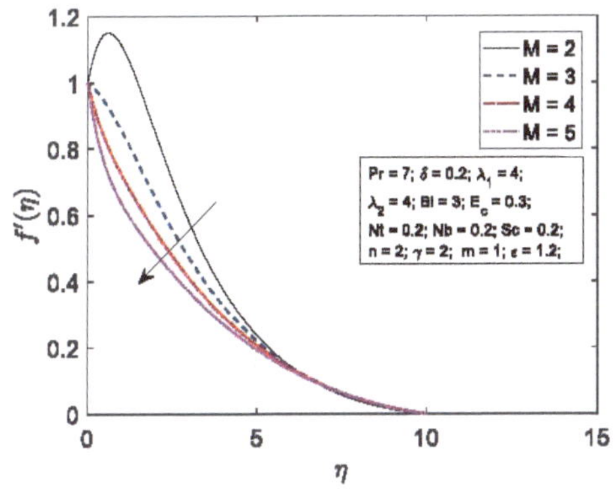

Fig. (6a). The effect of Magnetic parameter on momentum boundary layout.

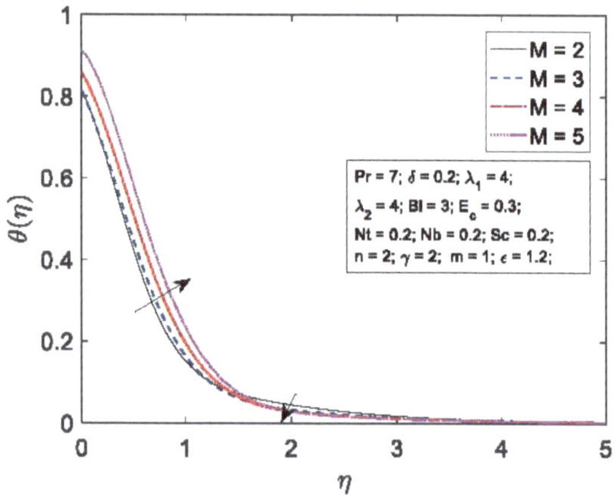

Fig. (6b). The effect of the magnetic parameter on the temperature boundary layout.

CONCLUSION

The problem of heat generation with the effect of the Brownian motion (Nb) and the thermophoretic parameter on a magnetohydrodynamic nanofluids flow through a non-linear stretching lamina with convective boundary condition in the appearance of higher-order chemical reaction has been investigated in this current study. A transformed set of non-linear ordinary differential equations was obtained and solved numerically by utilizing the Spectral Quasilinearization method. The impact of different physical parameters *Nb, Nt, Pr, Sc, and M* and heat-mass transportation characteristics are studied and analyzed in detail.

(i) The Brownian motion is inversely proportional to the velocity boundary layer.
(ii) With an increase in the Brownian motion parameter(N_b), temperature layout $\theta(\eta)$ is increased, on the other hand, the solutal boundary layer $\varphi(\eta)$ is decreased.
(iii) An increment of thermophoretic parameters leads to an enhancement of the velocity, temperature, and also in the concentration of the fluid.
(iv) We witness that the velocity considerably increases with the enhancement of the Schmidt number, but a reduction in fluid concentration.
(v) An increment of the Prandtl number causes a reduction in the temperature profile.
(vi) Magnetic parameter reduces the velocity but increases the temperature of the fluid profile.

ABBREVIATION

B Strength of uniform magnetic field

Bi Biot Number

C_p Specific heat of the fluid

C_∞ Ambient concentration far from sheet

D_B coefficient of Brownian motion

D_T coefficient of Thermophoretic diffusion

Ec Eckert Number

f Stream function

M Magnetic parameter

N_t Thermophoretic parameter

N_b Brownian motion parameter

ρ Density of the fluid

P_r Prandtl number

Sc Schmidt number

T_∞ Ambient temperature far from sheet

ν Kinematic coefficient of viscosity

σ Fluid conductivity

REFERENCES

[1] O.D. Makinde, and P. Sibanda, "Effects of chemical reaction on boundary layer flow past a vertical stretching surface in the presence of internal heat generation", *Int. J. Numer. Methods Heat Fluid Flow,* vol. 21, no. 6, pp. 779-792, 2011.
http://dx.doi.org/10.1108/09615531111148509

[2] T. Hayat, T. Muhammad, S.A. Shehzad, and A. Alsaedi, "Temperature and concentration stratification effects in mixed convection flow of an Oldroyd-B fluid with thermal radiation and chemical reaction", *PLoS One,* vol. 10, no. 6, p. e0127646, 2015.

http://dx.doi.org/10.1371/journal.pone.0127646 PMID: 26102200

[3] N.F. Sohut, A.S.A. Aziz, and Z.M. Ali, "Double stratification effects on boundary layer over a stretching cylinder with chemical reaction and heat generation", *In Journal of Physics: Conference Series, IOP Publishing,* vol. 890, no. 1, p. 012019, 2017.

[4] Y. Yongjae, "Importance of cooling rate dependence of thermoremanence in paleointensity determination", *J. Geophys. Res. Solid Earth,* vol. 116, no. B9, 2011.

[5] M.A.J.S. Van Boekel, "Kinetic modeling of food quality: A critical review", *Compr. Rev. Food Sci. Food Saf.,* vol. 7, no. 1, pp. 144-158, 2008.

http://dx.doi.org/10.1111/j.1541-4337.2007.00036.x

[6] P.O. Olanrewaju, O.T. Arulogun, and K. Adebimpe, "Internal heat generation effect on thermal boundary layer with a convective surface boundary condition", *Am. J. Fluid Dynamics,* vol. 2, no. 1, pp. 1-4, 2012.

http://dx.doi.org/10.5923/j.ajfd.20120201.01

[7] K. Bhattacharyya, "Effects of radiation and heat source/sink on unsteady MHD boundary layer flow and heat transfer over a shrinking sheet with suction/injection", *Front. Chem. Sci. Eng.,* vol. 5, no. 3, pp. 376-384, 2011.

http://dx.doi.org/10.1007/s11705-011-1121-0

[8] E. Abu-Nada, "Application of nanofluids for heat transfer enhancement of separated flows encountered in a backward facing step", *Int. J. Heat Fluid Flow,* vol. 29, no. 1, pp. 242-249, 2008.

http://dx.doi.org/10.1016/j.ijheatfluidflow.2007.07.001

[9] H.A. Mohammed, A.A. Al-Aswadi, H.I. Abu-Mulaweh, A.K. Hussein, and P.R. Kanna, "Mixed convection over a backward-facing step in a vertical duct using nanofluids—buoyancy opposing case", *J. Comput. Theor. Nanosci.,* vol. 11, no. 3, pp. 860-872, 2014.

http://dx.doi.org/10.1166/jctn.2014.3339

[10] A.K. Hussein, and S.H. Hussain, "Heatline visualization of natural convection heat transfer in an inclined wavy cavities filled with nanofluids and subjected to a discrete isoflux heating from its left sidewall", *Alex. Eng. J.,* vol. 55, no. 1, pp. 169-186, 2016.

http://dx.doi.org/10.1016/j.aej.2015.12.014

[11] T. Hayat, I. Ullah, A. Alsaedi, and M. Farooq, "MHD flow of Powell-Eyring nanofluid over a non-linear stretching sheet with variable thickness", *Results Phys.,* vol. 7, pp. 189-196, 2017.

http://dx.doi.org/10.1016/j.rinp.2016.12.008

[12] B. Mahanthesh, B.J. Gireesha, and R.S.R. Gorla, "Unsteady three-dimensional MHD flow of a nano Eyring-Powell fluid past a convectively heated stretching sheet in the presence of thermal radiation, viscous dissipation and Joule heating", *J. Assoc. Arab. Univ. Basic. Appl. Sci.,* vol. 23, no. 1, pp. 75-84, 2017.

http://dx.doi.org/10.1016/j.jaubas.2016.05.004

[13] S. Palani, B. Rushi Kumar, and P.K. Kameswaran, "Unsteady MHD flow of an UCM fluid over a stretching surface with higher order chemical reaction", *Ain Shams Eng. J.*, vol. 7, no. 1, pp. 399-408, 2016.

http://dx.doi.org/10.1016/j.asej.2015.11.021

[14] M. Dhlamini, H. Mondal, P. Sibanda, and S. Motsa, "Activation energy and entropy generation in viscous nanofluid with higher order chemically reacting species", *Int. J. Amb. Ene.*, vol. 43, no. 1, pp. 1495-1507, 2022.

http://dx.doi.org/10.1080/01430750.2019.1710564

[15] A. Sarkar, H. Mondal, and R. Nandkeolyar, "Powell-eyring fluid flow over a stretching surface with variable properties", *J. Nanofluids,* vol. 12, no. 1, pp. 47-54, 2023.

http://dx.doi.org/10.1166/jon.2023.1908

[16] H. Mondal, A. Mandal, and R. Tripathi, "Numerical investigation of the non-Newtonian power-law fluid with convective boundary conditions in a non-Darcy porous medium", *Waves Random Complex Media,* pp. 1-17, 2022.

http://dx.doi.org/10.1080/17455030.2022.2123966

[17] A. Sarkar, H. Mondal, and R. Nandkeolyar, "Effect of thermal radiation and nth order chemical reaction on non-Darcian mixed convective MHD nanofluid flow with non-uniform heat source/sink", *Int. J. Ambient Energy,* vol. 12, pp. 1-7, 2023.

[18] B. Uma, T.N. Swaminathan, R. Radhakrishnan, D.M. Eckmann, and P.S. Ayyaswamy, "Nanoparticle Brownian motion and hydrodynamic interactions in the presence of flow fields", *Phys. Fluids,* vol. 23, no. 7, p. 073602, 2011.

http://dx.doi.org/10.1063/1.3611026 PMID: 21918592

[19] R. Goyal, and R. Bhargava, "GFEM study of magnetohydrodynamics thermo-diffusive effect on nanofluid flow over power-law stretching sheet along with regression analysis", *arXiv preprint arXiv: 1708.05609,* 2017.

[20] M. Dhlamini, P.K. Kameswaran, P. Sibanda, S. Motsa, and H. Mondal, "Activation energy and binary chemical reaction effects in mixed convective nanofluid flow with convective boundary conditions", *J. Comput. Des. Eng.*, vol. 6, no. 2, pp. 149-158, 2019.

[21] F. Mabood, S.M. Ibrahim, and W.A. Khan, "Framing the features of Brownian motion and thermophoresis on radiative nanofluid flow past a rotating stretching sheet with magnetohydrodynamics", *Results Phys.*, vol. 6, pp. 1015-1023, 2016.

http://dx.doi.org/10.1016/j.rinp.2016.11.046

[22] P.M. Patil, S. Roy, and I. Pop, "Flow and heat transfer over a moving vertical plate in a parallel free stream: Role of internal heat generation absorption", *Chem. Eng. Commun.,* vol. 199, no. 5, pp. 658-672, 2012.

http://dx.doi.org/10.1080/00986445.2011.614978

SUBJECT INDEX

A

B

C

D

E

L

M

N

O

P

R

S

www.ingramcontent.com/pod-product-compliance
Lightning Source LLC
Chambersburg PA
CBHW041709210326
41598CB00007B/593